Exploring electronics

basic fundamentals

by

HOWARD H. GERRISH

Professor of Industrial Arts
Humboldt State College
Arcata, California

South Holland, Ill.

THE GOODHEART-WILLCOX CO., Inc.

Publishers

Instructor's
Guide and Answer Key
for use with the book

Exploring electronics

basic fundamentals

by
HOWARD H. GERRISH

South Holland, Ill.
THE GOODHEART-WILLCOX CO., Inc.
Publishers

TEACHING ELECTRICITY - ELECTRONICS

The educational foundations which support our present teaching programs have proved quite valid by extensive and continuing research. From the viewpoint of the Industrial Arts teacher, these programs offer the opportunity for a broad exploratory and discovery type of shop activity. All such programs should excite interest and motivation; they should be rich in guidance activities and be relevant to present practices in industry and technology. Your students will live and work in a technical society. A better understanding of the knowledge and skills required for this society will substantially contribute toward their success and happiness.

EXPLORING ELECTRICITY AND ELECTRONICS

As a teacher you may be interested in the basis on which the book, EXPLORING ELECTRONICS was written. Each of these criteria will be realized with varying degrees of emphasis as your students proceed through the instruction:

1. There must be as broad a coverage of the technical field as practical.

2. It must provide opportunity for discovery, study and experimentation and provide basic principles of electricity-electronics.

3. It should be enriched by the historical significance of science and invention.

4. The activities must be highly motivational.

5. On every occasion the program should encourage interrelationship and dependence on other school subjects such as reading, mathematics and science.

6. The instruction should be well organized and proceed step-by-step through established units of learning.

7. The program should be supported by excellent demonstrations involving both students and teacher.

8. There must be opportunities for original experimentation and creativity.

9. It must offer challenge to the better student yet provide training in basic skills and understandings for the average and below average student.

10. It must be rich in guidance.

11. It should encourage extra curricular activity.

12. It should provide construction and project building shop work to take full advantage of "learning by doing."

BUILDING PROJECTS

Although some experimental work will be done on the so-called "breadboard," the teacher must also realize that students want to make things. Project building activities can be very exciting and motivational. A good project should make a student ask, "Why does it work?"

Some of the best projects to use at this level of instruction will include some of these identities:

1. It should make a noise.

2. It should have visible action such as flashing lights or movement.

3. It must be low in cost.

4. It must be a device which the student can operate and adjust.

5. It should be a project which the student understands, but is not likely to be understood by nontechnical teachers or parents.

ELECTRONIC COMPONENTS
TO COMPLETE ALL EXPERIMENTS AND
PROJECTS IN EXPLORING ELECTRONICS

RESISTORS

1 — 100 Ω 2 w
2 — 200 Ω 1 w
1 — 1 KΩ .5 w
1 — 10 KΩ .5 w
1 — 1 MΩ .5 w
4 — 100 KΩ .5 w
3 — 470 KΩ .5 w
1 — 22 KΩ .5 w
1 — 47 Ω .5 w
1 — 4.7 KΩ .5 w

1 — Potentiometer 1 KΩ
1 — Potentiometer 10 KΩ
1 — Potentiometer 50 KΩ

2 ft. No. 20 Nichrome Wire

COILS AND TRANSFORMERS

1 — Ferri Loopstick Antenna Coil
1 — Argonne AR119
1 — Power Transformer 12.6 v CT
1 — Special Coil 1/2 in. Hollow Core

CAPACITORS

1 — Variable 20—365 pF
1 — Electrolytic 10 μF, 50 vdc
2 — Electrolytic 50 μF, 25 vdc
1 — Electrolytic 20 μF, 150 vdc
1 — Electrolytic 100 μF, 10 vdc
1 — Paper .1 μF
3 — Paper .5 μF
1 — Paper .001 μF
1 — Ceramic 02 μF
1 — Ceramic 100 pF

DIODES

1 — 1N34
1 — 1N295
2 — Diodes 500 mA, 400 v

TRANSISTORS

2 — 2N5449, NPN
1 — 2N107, PNP
1 — Thermistor

SWITCHES

2 — SPST
1 — DPST
2 — SPDT
1 — DPDT
1 — Rotary Switch
2 — Push Buttons
1 — Relay — Sigma 11F2300G
1 — Buzzer

INSTRUMENTS REQUIRED

DC Voltmeter 0—15 v
DC Ammeter 0—150 mA
Ohmmeter
Power Supply Variable 0—15 v
Audio Signal Generator
Oscilloscope

MISCELLANEOUS

2 — Round Magnets, 1/2 x 3 in.
4 — 1.5 v D Cells w/Holders
1 — B2M Sun Cell
1 — Clairex 505 Photocell
4 — 6.3 v Lamps No. 46
4 — Lamp Sockets
1 — PM Speaker 4 Ω
1 — Round Iron Core, 1/2 x 3 in.
3 — NE2 Glow Lamps
1 — Microphone
1 — Key
1 — AC Line Cord
1 — Earphone

SUPPLIES

1 — Zinc Electrode
1 — Copper Electrode
1 — Carbon Electrode
Salt
Copper Wire
Iron Wire
Iron Filings
Nails 4d
Compass

THE MOTIVATOR

A special group of components and parts which will complete most of the experiences and projects used in EXPLORING ELECTRONICS is manufactured by the Lab-Volt Division of Buck Engineering Co., Farmingdale, New Jersey 07727. This student-station-kit includes manuals, power supplies, meters and speaker in a convenient case.

The Motivator.

5

ANSWER KEY

UNIT 1, Page 16
Outside activities are self-explanatory.

UNIT 2, Page 24
1. Potential energy (See text, page 19).
   Electron (page 20).
   Negative Ion (page 21).
   Inductively charged (page 23).
2. The leaves will close.
3. Charge electroscope to known polarity. Observe leaf action when like or unlike charge is brought close to ball. If electroscope is charged negatively, a positive field will cause leaves to close.
4. To insulate center rod from cover on jar.
5. Each will be charged negatively and will repel each other.
6. Pith balls will attract each other.

UNIT 3, Page 33
1. This is an activity to teach color codes.
2. This introduces the use of ohmmeter and verifies color code value in question 1.
3. This is an experience in learning to use a voltmeter.
4. This experience teaches how to measure an unknown voltage.
5. This activity teaches how to connect and read an ammeter.

UNIT 4, Page 41
1. Perform as described.
2. P watts = I x E.   2 x 100 = 200 watts.
3. 2 x 500 watts = 1 KWH at .05 = 5 cents.
4. CURRENT $I = \dfrac{120 \text{ volts}}{20 \text{ ohms}} = 6$ AMPERES

   POWER $= I \times E = 6 \times 120 = 720$ watts

   also, POWER $= I^2R = 6^2 \times 20 = 720$ watts
5. 50 watts x 10 = 500 watts
   I in amperes x 120 volts = 500 watts
   $I = 4.16 +$ amperes
6. Excessive current, sparks, smoke, blow fuse or trip circuit breaker.
7. See Fig. 4-3, text page 38. Draw similar graph.

UNIT 5, Page 50
1. See Fig. 5-9, text page 46
2. No. Insufficient voltage and current.
3. Approximately 10 ohms cold.
4. Battery within ohmmeter causes lamp to glow.
5. Use Figs. 5-13 or 5-14, text page 48.
6. Use Fig. 5-16, text page 49.

UNIT 6, Page 66
1. 1.035 ohms.
2. Approximately 3 ohms.
3. Wire would heat. Lamp would be dim.
4. Removing any lamp will open circuit.
5. So total power loss by blown fuse or circuit breaker would not leave all parts of the house in darkness. Several circuits will cut down on wire size required and save cost.
6. No. So room would not be in total darkness with fuse failure.

7. Parallel lamps. Series circuits increase resistance. Parallel circuits decrease resistance.
8.
   A, B, C, and D.

9.

10.

UNIT 7, Page 79

1.

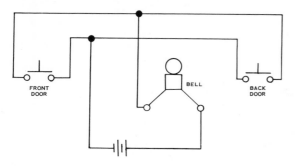

2. Glass does not affect magnetic field. Neither does copper. Magnet still picks up nails.
3. It proves that magnetic force depends upon current.

5. $I = mA$.
6. Draw wave form on oscilloscope.
7. Switch, move levers, turn valves, change electricity to in-line motion, door chime.

UNIT 8, Page 100

1. 1:10, 4:1, 1:4, 6:1, 16:1.
2. No transformer action. Primary winding would heat and destroy transformer.
3. Using a rectifier to change ac to dc.
4. See text Fig. 8-26, page 97. Use transformer with 24 volt center tapped transformer.
5. $I = .02A$. Since $.02 \times 500,000 = 10,000$ watts.
6. Student activity.
7. To step-down voltage for home use.
8. Full wave, less variation in output, higher frequency.
9. No transformer action.
10. The primary. It must carry 5 times more current.

UNIT 9, Page 113

1. The 22 K ohm resistor drops the voltage.
2. Circuit, Fig. 9-28, page 113. Use either a 90 volt B battery or build simple half-wave rectifier.
3. Circuit, Fig. 9-29, page 114. Use this for flashing eyes

for toy animals.
4. Circuit, Fig. 9-30, page 114. Very fascinating.
5. Circuit, Fig. 9-31, page 115.
6. A.

A and B = 100 KΩ		B and C = 0 Ω
C and D = 100 KΩ		E and F = 100 KΩ
F and D = 200 KΩ		F and C = 300 KΩ
F and A = 400 KΩ		

B. NE2 lamp will light. Follow directions.
C. Plug ac plug to 117 volt line. NE2 will light when probe and clip are connected together.

UNIT 10, Page 129

1. Circuit, Figs. 10-23 and 10-24, pages 128 and 129. Follow procedures.
2. Circuit, Fig. 10-25, page 130. Closing switch S applies forward bias to emitter-base junction and transistor will conduct.
3. Circuit, Fig. 10-26, page 130. Closing switch S applies forward bias to emitter-base junction. Transistor conducts and relay is activated by collector current.
4. No warm up – light weight – rugged – cheap – small – high gain – fast action.
5. Electrons.
6. Holes.
7. $A_v = 3000$.
8. Circuit, Fig. 10-27, page 130. Start with pure water. Distilled water is best. Decreased resistance of salt water allows base current to flow and therefore collector current will cause reacting to appear on meter.

UNIT 11, Page 145

1. Build circuit in Fig. 11-21, page 144. Removing C_2 should make reception weaker.
2. Build circuit Fig. 11-22, page 145. The transistor acts as an audio amplifier. Tuning is done with L_1 and C_1. It is a PNP transistor. C_2 is used to block base circuit from ground. Detection is accomplished by diode D.
3. Build circuit in Fig. 11-22, page 145. Note each block represents a part of circuit. Ground wires are omitted for simplification. The diode causes demodulation or detection. A PNP transistor is used. Yes, the circuit can be tuned by L_1. Connect up two transistor audio amplifier described in Fig. 10-23, page 146.

UNIT 12, Page 158

1.

Light	– I = .9 A	R = 122Ω
Toaster	– I = 5 A	R = 22Ω
Heater	– I = 9 A	R = 12.2Ω

2. Use circuit in Fig. 12-13, page 155.
3. Use thermistor and relay circuit.
4. Increased current due to lower CEMF. Heat equals I^2R in watts.
5. Use circuit in Fig. 12-11, page 154, with counter.
6. Use circuit in Fig. 12-10, page 153. Use other values of C if necessary.
7. Use circuit in Fig. 12-17, page 158. Lamp operates on 117 Vac. Rectifier supplies dc for photoresistive circuit and relay coil.

INTRODUCTION

EXPLORING ELECTRONICS is a first course in the interesting and challenging field of Electricity-Electronics. It will acquaint you with Basic Fundamentals of Electronics and their practical applications.

Interesting experiments will enable you to discover for yourself many electronic principles. Building simple projects and performing interesting activities will enable you to understand basic electronic circuits. Line drawings and photos will help you form a mental picture of the wonderful science of Electronics in action.

Extra color has been used in many of the illustrations to help make them meaningful, and to call attention to important points. Technical terms are defined as introduced, so you may proceed with confidence.

In recent years, the science of Electricity and Electronics has made many changes in our way of living. This enormous field of technology continues to expand and develop with new discoveries and applications. Electronics offers you a real challenge. It will be well for you to carefully explore this field and the rewarding careers it offers.

Howard H. Gerrish

CONTENTS

Contents

A complete Instruction Laboratory correlated to this text is available from the Lab-Volt Division, Buck Engineering Company, Farmingdale, New Jersey 07727.

The Laboratory (Motivator) includes manuals, component parts, connecting leads, power and metering equipment.

A Precision Tracking Dish. The antennas send and receive data from aircraft and transmit it to a computer system for processing. (Honeywell Electronic Data Processing)

Unit 1

THE ELECTRONIC AGE

ELECTRICITY AT HOME

The principles of electricity and electronics you will explore in this course find wide application in the home.

You can get an idea of the scope of electrical power used in the home from the list which follows:

LIGHT
Incandescent and fluorescent lighting.

FOOD PREPARATION AND COOKING
Electric ranges and ovens, electronic ovens, refrigerators, deep-freezers, ice cube makers and crushers, small appliances for mixing, blending, grinding, cutting and cooking, dishwashers and roasters, garbage disposals and can openers. See Fig. 1-1.

HEAT AND AIR CONDITIONING
Electric space heating, air conditioning, ventilation, automatic controls, electronic air filtering, humidifiers, and sump pumps.

PERSONAL CARE
Electric razors, toothbrushes, massagers, electric pads and blankets, vibrators, shoe polishers and brushes, hearing aids and hair dryers.

HOUSEHOLD CONVENIENCE
Radio controlled garage doors, floor polishers, intercommunications systems, photo light control, elevators, vacuum cleaners, and motorized curtains.

ENTERTAINMENT
Film and slide projectors, AM and FM radio, record changers, Hi-Fi monophonic (music produced by single output signal) and stereophonic (three-dimensional) sound systems, black and white and color TV, musical instruments, organs, guitars, recording equipment, and color organs.

Fig. 1-1. This appliance automatically cooks bacon and thin meats to perfection.

HOBBIES
Power machinery and tools, drills, sanders, paint sprayers, Ham radio, Fig. 1-2, CB radio, and rock tumblers.

GARDEN MAINTENANCE
Lawn mowers, cultivators, small tools, hedge trimmers, auto washers, and automatic sprinkler systems.

COMMUNICATIONS
Telephones — videophones.

SAFETY AND SECURITY
Burglar alarm systems, and fire alarm systems.

The list is far from complete. It does not list all items in the various categories and does not include electric typewriters, calculating machines and computers which will soon be in many homes. Nor does it include the family car which would remain in the garage if it were not for the storage battery.

Fig. 1-2. This student enjoys his own "radio shack" and chatting with friends. Ham radio is one of America's most respected hobbies. Many Hams continue their education in the fields of electric and electronic technology.
(American Radio Relay League)

ELECTRICITY AND ELECTRONICS IN INDUSTRY

In industry, as well as in the home, the electric motor can well be called "man's servant." From a tiny motor to drive a delicate instrument to giant motors capable of developing hundreds of horsepower, the motor is ready and able to perform at the flip of a switch. It is quiet and efficient, requires little attention and is relatively easy to connect to the machine and power source. Motors provide a very economical method of converting electrical energy into mechanical energy.

The generation of electric power is also an industry in its own right. In past years, most power generators were turned by steam turbines. Many of these plants are still in existence, but the dwindling supply of coal has encouraged engineers to search for other sources of power. In some areas hydroelectric plants use the energy of water stored behind dams as it falls to operate huge turbines and generators.

Fig. 1-3. Pressure test of the gas thermal steam wells in Napa County, California. These wells are connected to turbine-generator units.

Fig. 1-3 is not a picture of Yellowstone National Park. The steam wells shown in the photo, located in Napa County, California, are the source of steam power used to operate a geothermal (heat obtained from the earth) power plant.

Since the first explosion of an atomic bomb, scientists have searched for ways and means of converting the enormous energy of the atom into peaceful uses. The atomic power plant is one such accomplishment. The nuclear reactor is used to heat steam to drive the turbine-generators. The control room of an atomic power plant is illustrated in Fig. 1-4. Atomic energy may well supply the answer to our increasing needs for electric power.

You will be interested in Fig. 1-5 which shows the largest nuclear power electric generator built to date.

Electronics in industry appear more in instrumentation and control of machinery

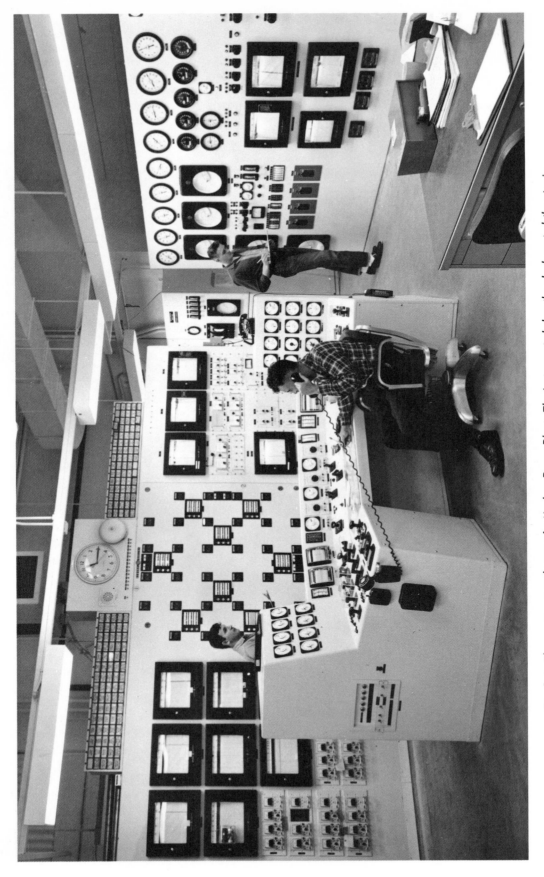

Fig. 1-4. The reactor control room of a Nuclear Power Plant. The instruments below the clock control the atomic reactor power levels. New uses of atomic energy are opening up almost unlimited opportunities for young people. (Pacific Gas and Electric Co.)

and processes. Manufacturing processes which once required close and visual supervision may now be observed and controlled from remote locations. This alone increases safety and economy in manufacturing.

The conversion of electrical energy to heat energy will also find broad usage in industry. All kinds of heating requirements are met, from heating a small office to a high temperature electric furnace for melting steel. One less well known application of heating is inductive heating. It is used extensively in heat treating and tempering metals. The part to be heated is placed in a coil. High currents are induced in the material to be heated by transformer action and the part is heated by its own internal molecular friction.

fabrication of farm machinery. From building automobiles to battleships, the electric arc welder serves industry by increased production speed and economy.

Fig. 1-6. Electric welder being used to assemble farm machinery (Westinghouse)

Another common industrial use of electrical energy which is now finding its way into your home, is ultrasonic cleaning and washing. In this application the parts to be washed are not sloshed around in soapy water, but are placed in a tank filled with the cleaning liquid. The liquid is caused to vibrate by an electronic generator at ultrasonic frequencies.

ELECTRICITY AND ELECTRONICS IN TRANSPORTATION

Transportation by land, by air or by sea uses electricity for power, direction and control. Electric trains used in many areas are fast and clean, and provide services and comforts that were only dreams a few years ago. The increased railroad traffic demands precise control and scheduling. In this capacity, electronics steps forward with electronic observation and control; computers provide fast computations and decisions. These are requirements of this Electronic Age. Fig. 1-7 shows the centralized traffic control for a complete railroad system.

Fig. 1-5. The largest nuclear power electric generator built to date. Nuclear power plants may become the source of electrical power for our future needs. (Westinghouse)

Electricity also is the servant of electric welding, the process in which metal parts are joined together by fusion (combining metals by melting together) of the metals. The welder, shown in Fig. 1-6, is used in the

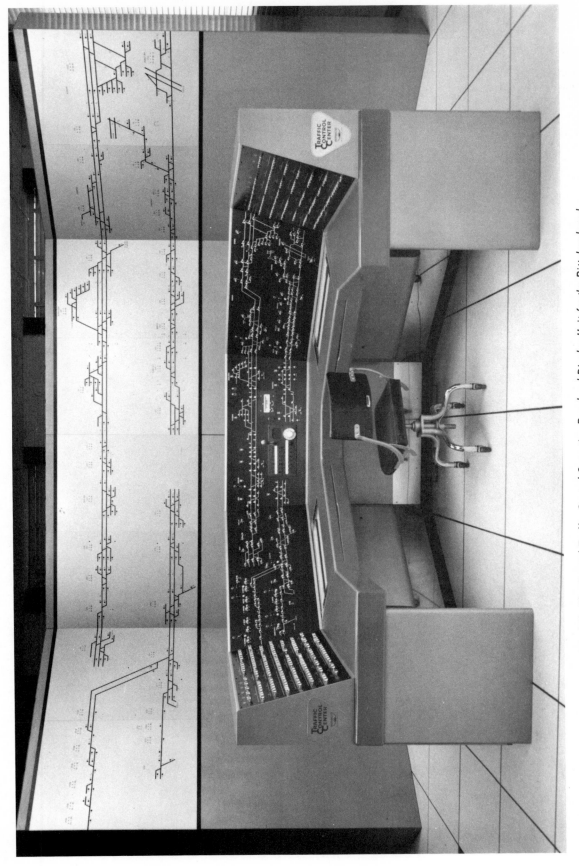

Fig. 1-7. The Central Traffic Control Operation Panel and Display Unit for the Pittsburgh and Lake Erie Railroad. Movement of all trains can be observed and controlled by this system. (Westinghouse Air Brake Co.)

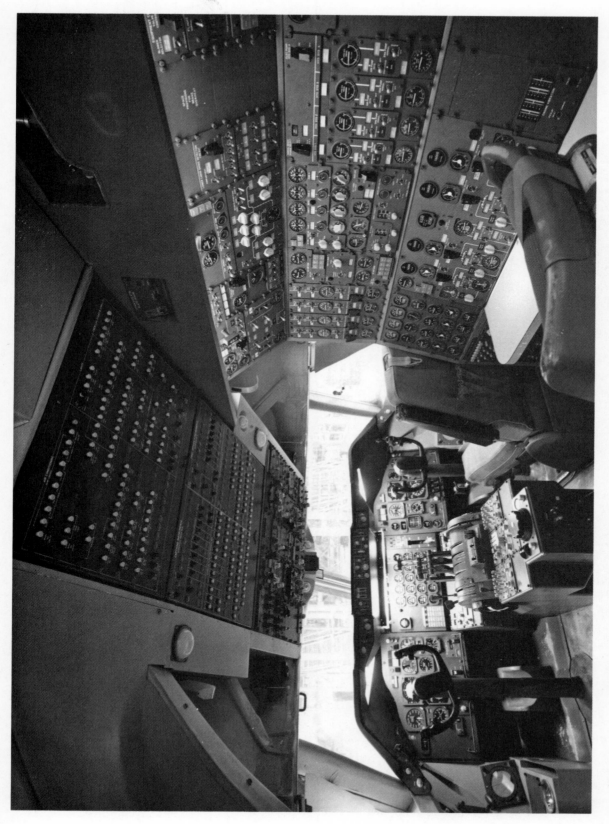

Fig. 1-8. The flight deck of the Boeing 747 airplane shows the instruments and controls for the pilots and flight engineers. The instruments monitor the operation and performance of all systems in the aircraft. Specialists in electronics are required to maintain and operate this equipment.

Ships at sea no longer need to depend on the compass for navigation. The pilot house is now an electronic technician's dream. Immediate voice communication with the home port is provided by radio. Radar is used to survey the horizon for approaching ships and dangerous reefs. The ship's captain can "see in the dark." Electronic depth sounders immediately indicate the distance to the ocean floor. Direction finders and radio compasses can instantly show the position of ship or aircraft.

Electricity provides the power aboard ship for hoists, winches, cranes and elevators, and many ships are powered by electricity. On naval ships, electronics takes over the control of "fire power." Guns are pointed, trained, and fired, all by electricity and electronic circuits.

Probably the most sophisticated and complex system of electronic circuits is used to control and direct our aircraft. Look at Fig. 1-8. This gives you an idea of what it takes to fly Boeing 747 planes. Needless to say, extremely skilled pilots and engineers are required for its safe and efficient operation. On the ground, highly skilled and trained technicians must inspect and maintain all aircraft at peak performance.

ELECTRICITY AND ELECTRONICS IN THE COMMUNICATIONS INDUSTRY

Communicating by radio waves is an important division of electronics. At home, in industry, in transportation, and in space travel, communication by radio waves is a part of each system. In television the picture and sound require separate transmitters. Fig. 1-9 shows a typical television studio. "Video tape recording" has changed the nature of TV broadcasting in recent years. Studio programs can be taped at a convenient time, and the tape held until such time as the program is needed. By video tape recording we enjoy the instant playback of athletic teams in action. It is also a rich source for summer "reruns." A video tape recorder is shown in Fig. 1-10.

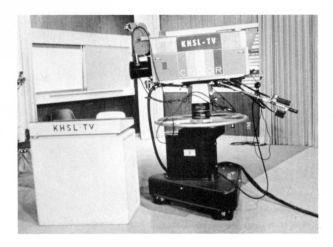

Fig. 1-9. TV studio. Many technicians and engineers are needed in commercial broadcasting. A First Class FCC License is usually required.
(KHSL, Golden Empire Broadcasting Co.)

The telephone which has served so well through the years, provides a dependable and inexpensive method of electronic communications. Great improvements are being made continuously in the telephone industries. Telephone lines which reach around the world use wires and microwave radio. Others bounce signals from communications satellites. Such accomplishments defy the imagination as to the limits of possibility.

Fig. 1-10. Broadcast video tape recorder. Suitable for commercial and educational TV stations. (Ampex Corp.)

Now the Picturephone has become practical. With this, you can SEE and HEAR during your telephone conversations.

ELECTRONICS IN MEDICINE

The science of electronics serves well in the hands of doctors and surgeons in the hospitals of the world. Many of the instruments devised provide continuous observation, monitoring and recording the condition of patients. Alarm systems are provided to call nurses and doctors in emergency cases. Information such as pulse rate, respiration, blood pressure, and body temperature is available instantly. Other instruments are designed to use electronic circuitry to aid in body functions. Such is the "Gift of Life" made possible by the Pacemaker which provides an electrical stimulus to cardiac (heart disease) patients.

Fig. 1-11. The Cathode Ray Tube (like a television screen) provides up-to-date information stored by the hospital computer. The Toy Dog? Just to make the hospital friendly to a child. (Honeywell, Inc.)

Another phase of medical care includes scheduling of patients and recording their treatments. Information about the patient is stored in a computer memory and may be shown immediately on a cathode ray tube, Fig. 1-11.

Your study of the preceding pages will give meaning and understanding to the following Units. Electronics is an extremely interesting study and a challenge to all youths who live in the Age of Electronics.

OUTSIDE ACTIVITIES

1. Plan a field trip to a local industry and report on the types of electrical machinery and equipment in use. Identify the electrical effect produced such as heat, light, sound or motion.
2. From your reading or field trip, select one particular machine or process involving electricity and make a verbal detailed explanation to your class. This will probably require some study and research.
3. Make an appointment with an electrician, technician or engineer and ask him about his vocation or profession. Prepare the questions you wish to ask before the interview. Make a verbal report to your class.
4. Make a list of all the books and their authors on electricity and electronics which are in your school library. Post the list on the shop bulletin board.
5. Arrange for an electrician, technician or engineer to visit your class and talk about his vocation. Before his arrival, prepare questions for him to answer.
6. Count and list the purpose of all the electric motors used in your home.
7. Prepare a list and if possible secure samples of magazines on electricity and electronics available at your local news stand. Keep the magazines for your fellow students to look over. You may wish to subscribe to one or more.
8. Is there a Ham radio operator in your class or do you know of a friend who is a Ham? Ask him or her to talk about the hobby and demonstrate his radio equipment (gear).
9. Ask a fellow student or teacher who is interested in Hi-Fi Amplifiers to talk and demonstrate his Hi-Fi equipment. Have him make some tape recordings of your class discussions or talks by engineers and technicians.

RULES OF SAFETY

1. Acquire the safety habit.
2. Always think first and observe caution.
3. Keep your shop and bench clean and in order.
4. Use the right tool for the right job.
5. Operate machines only after proper instruction.
6. Wear goggles, face shields and protective clothing when required.
7. Don't horse around. No jokes.
8. Don't interrupt or talk to a fellow worker when he is operating a machine.
9. Handle all materials carefully.
10. Know where the First Aid Kit is located and always get first aid for any injury.
11. Use common sense.

TUNING FORK

ACCUTRON

The transistorized wrist timepiece uses a tiny tuning fork as its frequency standard. The watch is guaranteed to the on-the-wrist accuracy within 60 seconds per month. It is powered by a mercury-oxide power cell about the size of an aspirin tablet. Illustrated is a model with a transparent dial which permits the owner to observe its operation. (Bulova Watch Co.)

During this earth test of a communications satellite, it is spun at 90 rpm to test its pointing accuracy. As a satellite spins in space, its sensors keep track of the Earth horizon and the position of the sun. (Philco—Ford Corp.)

Fig. 2-1. Power companies create lakes in the mountains by constructing dams. The potential energy stored in the lake is used to produce electricity. Lakes also provide recreation and scenic beauty. Some are also used for water supply, irrigation and flood control.   (Pacific Gas and Electric Co.)

Unit 2

WHAT IS ELECTRICITY?

Electricity is a kind of ENERGY. Energy may be defined as the capacity or ability to do work.

POTENTIAL ENERGY

POTENTIAL ENERGY is energy at rest. Let's see what this means. A scientist describes a body of water, such as the mountain lake shown in Fig. 2-1, as possessing POTENTIAL ENERGY. This means the water in the lake has the capacity to do work. When released to flow into a power plant, energy from the falling water operates a turbine to generate electricity.

Why is the water falling? It falls because of gravity. The laws of gravity were formu- lated by Sir Isaac Newton in the 17th century.

Let's find out more about the POTENTIAL ENERGY stored in the mountain lake. How do we measure this POTENTIAL? An ex- ample will help you understand the measure- ment of energy. If you were to lift a ten pound weight to a height of three feet, the weight would acquire thirty FOOT-POUNDS of POTENTIAL ENERGY. (3 ft. x 10 lbs. = 30 ft. lbs.) If you dropped the weight (not on your foot) the stored ENERGY could do THIRTY FOOT-POUNDS OF WORK. PO- TENTIAL ENERGY can be further defined as ENERGY STORED AS A RESULT OF POSITION. Our lake has potential energy due to its height or elevation above the power plant.

OUR WORLD IS MADE OF MATTER

MATTER is defined as anything which takes up room or occupies space. Matter may be in the form of a solid, liquid or gas. All matter is made up of the basic elements of nature. Each element has been given a name. Familiar elements include gold, silver, iron, oxygen, hydrogen and carbon. There have been over one hundred elements discovered and identified. Matter consists of the basic elements either in pure form or in mixtures and compounds. For example: water consists of two parts of hydrogen and one part of oxygen (H_2O); salt consists of one part sodium and one part chlorine (NaCl).

ATOMS - PROTONS AND ELECTRONS

Elements may be divided into parts. The smallest unit into which an element may be divided without losing its identifying characteristics is called a MOLECULE. A molecule consists of one or more like atoms in an element or two or more different atoms in compound.

An atom is the smallest part of an element having all of the properties of the element. ATOMS consist of electrically charged particles. Some particles have a positive charge (+ charge) and are called PROTONS. Some have a negative charge (- charge) and are named ELECTRONS. Others have neither a positive nor negative charge and are called NEUTRONS. We are concerned in particular with the ELECTRON.

A MODEL OF THE ATOM

A simplified explanation of the structure of an atom is shown in Fig. 2-2. For our discussion we may consider that the center or NUCLEUS of the ATOM is made up of PROTONS and NEUTRONS. Surrounding the nucleus in specific rings or orbits are found the ELECTRONS. In the illustration the atomic models of carbon (C) and silicon (Si) are shown. Notice that a carbon atom has six protons or positive charges in the nucleus and two orbits containing six negative

charges or ELECTRONS. In the silicon atom there are fourteen positive charges or protons in the nucleus and three orbits containing fourteen negative charges or ELECTRONS.

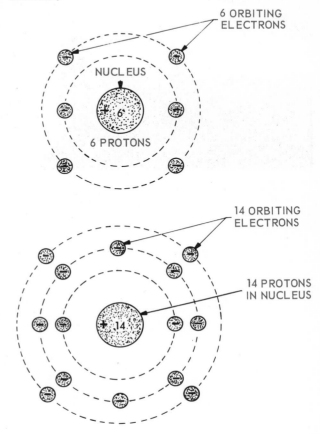

Fig. 2-2. Above. The Carbon atom has six **ELECTRONS** in its outer rings or orbits. Its atomic number is 6. Below. The Silicon atom has fourteen **ELECTRONS** in its outer rings or orbits. Its atomic number is 14.

GREEDY OR GENEROUS

These are peculiar words to use in electricity. Why are they used? Looking back to Fig. 2-2 you will notice that in both atomic models the NUMBER OF PROTONS IS EQUAL TO THE NUMBER OF ELECTRONS. Since they are equal in number and opposite in charge, the atom has NO ELECTRICAL CHARGE. The atom is therefore NEUTRAL.

Study the location of the PROTONS for a moment. They are gathered together in the nucleus and are difficult to reach. However,

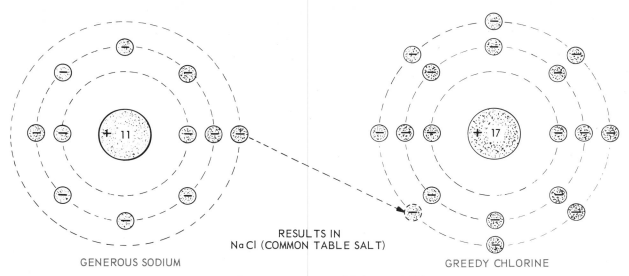

RESULTS IN
NaCl (COMMON TABLE SALT)

GENEROUS SODIUM

GREEDY CHLORINE

Fig. 2-3. The atomic structure of Sodium and Chlorine

the electrons are moving about in circles and are rather easily disturbed. If some electrons could be removed from their orbits, then the NUMBER OF PROTONS WOULD BE GREATER THAN THE NUMBER OF ELECTRONS. The PROTONS would be in the majority and the atom would indicate a POSITIVE CHARGE. A positive ION has been made.

Now look at the opposite situation. Suppose an atom is greedy and would capture an electron from a closely associated but generous atom. Then the atoms would have MORE ELECTRONS THAN POSITIVE PROTONS. The ELECTRONS would be in the majority. The atom would indicate a NEGATIVE CHARGE. A negative ION has been produced. In Fig. 2-3 the chemical reaction to form a simple grain of table salt is shown. Which one is generous and which is greedy? Chemical changes are assumed to occur in this manner.

THE LAW OF ELECTRIC CHARGES

Some interesting experiments may be performed to demonstrate the presence of an electrical charge and how these charges behave. It is suggested that you set up these experiments and satisfy your own curiosity. If you understand the Laws of Electrostatic Charges, you are well on your way toward

an exciting study of electricity. ELECTRO-STATICS? That means "electricity at rest." Later on you will study "electricity on the move."

You have probably experienced the generation of an electrostatic charge with surprising results. If the family car has the usual type of seat covers and you slide over these covers, you may get a sudden shock when you touch the metal door handle. Have you a family pussycat? Pat your kitty and be rewarded with a friendly purr. Now bring your hand in a backwards direction just above the kitten's fur. Note how his fur will stand on end. Sometimes you will hear the crackling of electric sparks. (Warning - this

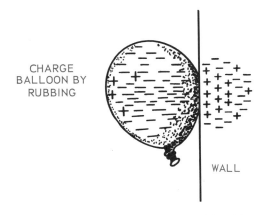

CHARGE
BALLOON BY
RUBBING

WALL

Fig. 2-4. A balloon is attracted because unlike charges are produced by the charged balloon.

Fig. 2-5. The construction of an **ELECTROSCOPE**. The glass jar protects the delicate foil strips from air movement.

may not be appreciated by the cat.) Take a toy balloon, blow it up and tie its end. Rub the balloon vigorously with a piece of wool cloth. Now touch the balloon to the wall of your room. It mysteriously sticks to the wall. An explanation of this mystery is given in Fig. 2-4. Friction produces a negative charge on the balloon. As the balloon approaches the wall, it repels the electrons in the wall which leaves a positive area. This demonstrates the LAW OF CHARGES:

LIKE CHARGES REPEL EACH OTHER.
UNLIKE CHARGES ATTRACT EACH OTHER.

The fact that one electron repels another electron is a fundamental law in electricity.

BUILD AN ELECTROSCOPE AND EXPERIMENT

Definite conclusions and proof of the Laws of ELECTROSTATICS may be gained by building a simple ELECTROSCOPE. This instrument detects the presence and polarity (negative or positive) of an electrostatic field. Any charged body will have an INVISIBLE FIELD OF FORCE surrounding it. This will be proved by the electroscope.

Constructional details on an electroscope are given in Fig. 2-5. Any glass jar may be used. Bend a twelve inch length of heavy copper wire to form the center rod. On the hooked ends, hang pieces of light, thin tinfoil. The foil should be trimmed to about three eighths of an inch wide and one inch long. Wrap the bent copper wire with plastic tape for insulation. Now you are ready to try it out.

In the physics laboratory two convenient materials are used to produce an electrostatic charge. A hard rubber or vulcanite rod rubbed with a wool cloth will charge the rod negatively. Electrons will be transferred from the wool to the rod by the rubbing action. A glass rod rubbed by a piece of silk cloth will charge the glass rod positively. Electrons will leave the glass and move to the silk by the rubbing action. If vulcanite and glass rods are not on hand, borrow them from the science department. You might also experiment with other kinds of material. Try your plastic fountain pen.

EXPERIENCE 1. Detect an electrostatic charge. Rub the vulcanite rod vigorously with the wool cloth to build up a negative charge. Bring the rod close to, BUT NOT TOUCHING the electroscope rod. Note that the foil leaves in the jar move outward and try to get away from each other. See Fig. 2-6.

WHY? Existing around the negatively charged vulcanite rod is an invisible electrostatic field or force. When brought close to the top end of the electroscope, the electrons at the top are driven downward to the foil leaves. Both leaves become negative and repel each other, therefore spreading apart. LIKE CHARGES REPEL. Remove the vulcanite rod and note that the leaves return to their normal position.

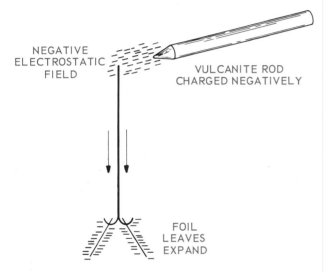

NEGATIVE ELECTROSTATIC FIELD

VULCANITE ROD CHARGED NEGATIVELY

FOIL LEAVES EXPAND

Fig. 2-6. Electrostatic field around charged rod drives electrons on electroscope rod to the foil leaves. Both leaves become negative and REPEL EACH OTHER.

EXPERIENCE 2. Charge the electroscope INDUCTIVELY. A new word, INDUCTIVELY means: "by the influence of, or to cause it to become." Once again charge the vulcanite rod and bring it close to the electroscope. The leaves will expand. Now while holding the rod in this position touch the top of the electroscope with your finger. The leaves will close. Now REMOVE YOUR FINGER AND THEN THE ROD. The leaves expand again and remain in this position. The electroscope is charged POSITIVELY.

WHY? The first action is the same as in the first experiment. However, when you touched the electroscope, electrons repelled by influence of the negative rod were also forced into your body through your finger. You removed your finger and when the rod was also removed the electroscope had no means of regaining the electrons which were forced into your body. Therefore the electroscope is left with fewer electrons. It is electrically out-of-balance and is positive. Both leaves are now positive and repel each other. LIKE CHARGES REPEL. To further prove this, touch the electroscope with your finger. The leaves close and remain closed. The ELECTRONS from your body were returned to the electroscope and made it neutral.

EXPERIENCE 3. CHARGE THE ELECTROSCOPE BY CONTACT. Only a weak charge on the vulcanite rod is required, so rub it lightly. Touch the electroscope with the charged rod and then remove it. The leaves expand and remain expanded. The electroscope is charged NEGATIVELY.

WHY? Since the rod is negative with an excess of electrons and the electroscope is neutral, the total charge on the rod shares its electrons with the electroscope when contact is made. Of course, when the rod is removed, the electroscope cannot rid itself of these electrons and therefore remains charged NEGATIVELY.

EXPERIENCE 4. Repeat EXPERIENCES 1, 2 and 3 using the glass rod and silk.

FORWARD STEPS TO UNDERSTANDING ELECTRICITY-ELECTRONICS

1. Electricity is a form of ENERGY.
2. Potential energy is energy at rest and has the capacity to do work.
3. Gravity is a universal law of nature.
4. Potential energy is energy stored as a result of position.
5. Matter is anything which occupies space.
6. All matter is made up of the basic elements of nature.
7. A molecule is a unit of matter.
8. Molecules are made of small particles called atoms. Each atom has its own

characteristics.

9. A PROTON is a positively charged particle.
10. An ELECTRON is a negatively charged particle.
11. The nucleus of the atom usually contains protons and neutrons.
12. An electrically unbalanced atom becomes an ION.
13. Like charges repel each other.
14. Unlike charges attract each other.
15. Any charged object is surrounded by an invisible electrostatic field of force.

TEST YOUR KNOWLEDGE - UNIT 2

Write your answers to these questions on a separate sheet of paper. Do not write in this book.

1. Write out in your own words the meaning of:
   Potential Energy
   Electron
   Negative Ion
   Inductively Charged
2. An electroscope has been charged negatively. What will happen if a positive rod is brought close to it?
3. How can the electroscope be used to discover if a body is charged negatively or positively?
4. In building the electroscope, why is the center wire wrapped with tape?
5. In Fig. 2-7, two pith balls are suspended on threads. What will happen if both balls are charged by contact with a charged vulcanite rod?
6. In Fig. 2-7, what will happen if ball A is charged negatively and ball B positively? Explain.

RULES OF SAFETY

SHOCKS ARE DANGEROUS

Electricity is a "shocking affair." You can get shocked when working around electrical equipment. Getting shocked is usually due to CARELESSNESS. Always THINK BEFORE ACTING.

HAND TO HAND INCREASES DANGER

In electricity, if you take hold of a live circuit with both hands, current flows across your chest and heart. This is not a safe practice. The TV technician, working with high voltages, generally keeps one hand in his pocket. It may be inconvenient, but it is safer to work with one hand only.

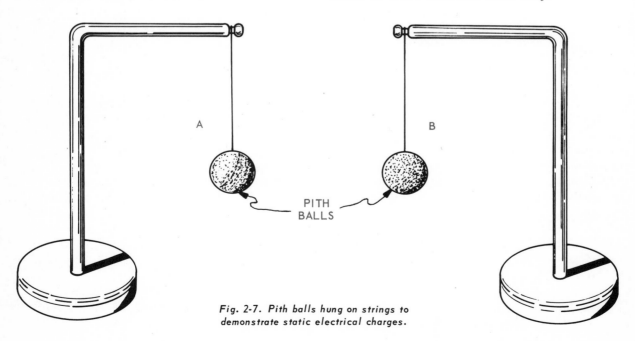

A

B

PITH
BALLS

Fig. 2-7. Pith balls hung on strings to demonstrate static electrical charges.

Unit 3

THE LANGUAGE OF ELECTRICITY

WHAT IS A VOLT?

In Unit 2 you read about potential energy. This was defined as energy at rest. Potential energy has the capacity to do work. Studying Unit 2, you performed experiments to prove that atoms can become negative or positive IONS by the transfer of electrons from generous to greedy atoms.

NEGATIVE   POSITIVE

A   —   +   B

MORE ELECTRONS THAN PROTONS   LESS ELECTRONS THAN PROTONS

Fig. 3-1. Charged bodies or terminals have a potential difference.

Let's look at Fig. 3-1. Here we have two balls or terminals. One has been charged in a negative direction; the other has been charged in a positive direction. One has many MORE electrons than protons; the other has many LESS electrons than protons. We say that the body of the terminal on the LEFT (A) is NEGATIVE, this is marked with a minus sign (-). The terminal on the RIGHT (B) is POSITIVE and is marked with a positive sign (+).

A source of potential energy is created by the flow of electrons in a conductor between the terminals. In mechanics this potential energy is measured in foot-pounds. In electricity the potential is measured in VOLTS. This term was so named to honor

Alessandro Volta, an Italian scientist. The letter symbol for volt is E or V. When the electrician speaks of VOLTS or VOLTAGE, he is talking about a POTENTIAL DIFFERENCE between two terminals. It may also be referred to as a FORCE, an ELECTROMOTIVE FORCE or EMF. Because this voltage or potential difference or EMF exists, the source of energy has the capacity to do work. The term VOLT probably is new to you. At home the lights and appliances operate on 117 volts. The family car has a 12 volt battery. Your transistor radio may use a 9 volt battery. Your flashlight may use two or three 1.5 volt cells. The exact value of a volt as we measure it, has been established by international standards, and is the same anywhere in the world.

IMPORTANT. VOLTAGE IS ALWAYS MEASURED BETWEEN TWO POINTS IN AN ELECTRICAL CIRCUIT. THE VOLTAGE IS THE DIFFERENCE BETWEEN THE TWO POINTS. Voltage is static nonmoving pressure only. At home the water pressure at the faucet in the kitchen sink may be 20 lbs. per square inch. NO WATER IS FLOWING. Yet the pressure is always available to force out water if you turn on the faucet.

HOW TO MEASURE VOLTAGE

The instrument used to measure voltage is called a VOLTMETER. It has two test leads. The test leads are usually red and black. The RED lead is the POSITIVE LEAD and is connected to the positive terminal (highest potential point) in the circuit. The

BLACK lead is the NEGATIVE LEAD and must be connected to the negative terminal (lowest potential point in the circuit). FAILURE TO CONNECT A VOLTMETER CORRECTLY MAY DESTROY THE METER. In Fig. 3-2, the symbol for a voltmeter is used and the correct connections are made.

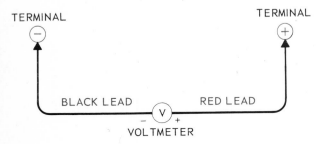

Fig. 3-2. *A voltmeter must always be connected with black negative lead to negative and red positive lead to positive.*

The value of the voltage or potential difference measured is read on the voltmeter scale. A simple voltmeter scale is illustrated in Fig. 3-3. Note that the scale reads from zero to 15 volts. At one third of the scale there is a 5 volt mark; at two thirds of the scale a 10 volt mark. Note also that in between 0 and 5 and 10 and 15 are shorter indicating marks which have no numbers. Each of the longer marks has a value of ONE VOLT. Between each of the one volt marks are shorter marks. Each of these marks has a value of .5 or one-half volt.

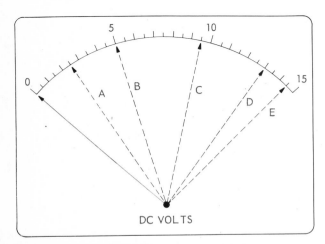

Fig. 3-3. **VOLTMETER SCALE FOR 15 VOLTS.** *Pointer in Position* A = 2.5v; *Position* B = 5v; *Position* C = 9.5v; *Position* D = 13v; *Position* E = 14.5v.

This meter scale is so arranged that you may measure any voltage quite accurately between zero and 15 volts in steps of .5 volts. Study each position of the indicating pointer in the illustration.

NOTE: A meter such as this example can ONLY measure up to 15 volts. If you attempt to measure a voltage higher than 15 volts, it will destroy the meter.

WRONG CONNECTIONS DESTROY METERS

A voltmeter must be connected with the right polarity. Black or negative lead to the negative or ground point in a circuit. Red or positive lead to the positive or high voltage point in the circuit. Failure to do so causes the needle to move in the wrong direction.

MULTIRANGE VOLTMETER

Quite often you will need to measure voltages much higher than 15 volts. For this purpose, a multirange voltmeter is used. This has two or more meter scales and a RANGE SWITCH. A typical scale,

Fig. 3-4. *The scale of a* **MULTIRANGE VOLTMETER** *with* **RANGE SWITCH. NEEDLE** *is shown in positions A, B, C and D.*

Fig. 3-4, has a single scale with two sets of values. The values to use is determined by the setting of the RANGE SWITCH. The range, of course, will be determined by how high a voltage you want to measure.

Let's look at the scale and each set of values. SCALE 0 to 25, has major divisions marked at 5, 10, 15, 20 and 25. Then each major division is further divided into ten parts. Each mark has a value of .5.

SCALE 0 TO 5, has major divisions marked at 1, 2, 3, 4 and 5. Then each major division is divided into ten smaller parts. Each mark has a value of .1.

WHICH SCALE VALUE TO USE? Use the scale value divided to match the RANGE SWITCH POSITION. Study the table, Fig. 3-5.

RANGE SWITCH POSITION	SCALE TO READ
5v	0–5
25v	0–25
50v	0–5
250v	0–25
500v	0–5

Fig. 3-5. Scale to read for range switch position.

Get the idea? It requires a little thought. To get some practice, the indicating needle in Fig. 3-4 is shown in positions A, B, C and D. The values of the indicated voltage for each range switch position appear in the table, Fig. 3-6. It is suggested that you read the meter first and then compare your readings with the answers given in the table.

RANGE SWITCH POSITION		VOLTAGE VALUE AT INDICATOR POSITION			
	SCALE	A	B	C	D
5v	0–5	1.4v	2.5v	3.8v	4.5v
25v	0–25	7v	12.5v	19v	22.5v
50v	0–5	14v	25v	38v	45v
250v	0–25	70v	125v	190v	225v
500v	0–5	140v	250v	380v	450v

Fig. 3-6. Answers to voltmeter reading experiences.

USING PREFIXES

In our studies of electricity we add a prefix or letters at the beginning of a word

to change its meaning. Such is the case with VOLTS. Note these prefixes and their values:

1 MEGAVOLT	=	1 MILLION VOLTS (1,000,000 VOLTS)
1 KILOVOLT	=	1 THOUSAND VOLTS (1,000 VOLTS)
1 VOLT	=	1 VOLT
1 MILLIVOLT	=	1/1000 VOLT (.001 VOLT)
1 MICROVOLT	=	1/1000000 VOLT (.000001 VOLT)

Frequently the Greek letter (mu) μ is used for "micro." These prefixes will also be used with other units of measurements.

USE CORRECT SCALES FOR METER SAFETY

If you need to measure 100 volts, be sure to turn the "range selection switch" to its proper position. When measuring UNKNOWN VOLTAGES always start with the highest range. You can switch down to a lower voltage range if necessary.

ELECTRICITY ON THE MOVE

Thus far we have kept the electrons which created a potential difference or voltage in prison, so to speak. They have had no freedom to run about. Yet we did state that a potential voltage has the capacity to do work.

The two charged bodies or terminals have been redrawn in Fig. 3-7 with one very significant change. A copper wire has been connected between them. This copper wire is a CONDUCTING PATH and electrons will move from negative to positive until the charge on ball A is equal to the charge on ball B and NO POTENTIAL DIFFERENCE exists. This movement of electrons is called CURRENT. Its letter symbol is I. Current is measured in AMPERES. It is this current, as you shall see, that does the WORK.

Fig. 3-7. The difference in potential causes a current to flow when a conducting wire is connected between the terminals.

NOTE: Some texts consider electricity as flowing from positive to negative and it is referred to as "conventional current flow." In this text current will always be considered as "electron flow" and will move from NEGATIVE TO POSITIVE.

POTENTIAL DIFFERENCE REQUIRED

Let's refer back to the mountain lake in Unit 2. What would happen if the lake had no way to replace the water? Soon the lake would be dry. The potential energy provided by the water stored in the lake would be gone. The same condition results in Fig. 3-7. Current flows as long as a potential difference exists. If it is not RECHARGED we have a DEAD SOURCE OF ENERGY. NO CURRENT. NO WORK CAN BE DONE. THERE MUST BE A POTENTIAL DIFFERENCE FOR CURRENT TO FLOW.

SAFETY FOR METERS

Meters are "the seeing eyes" of the technician. They are delicate and precise instruments. This means careful handling. Don't slam one down on the bench. Be gentle. Don't drop it on the floor. Meters should be kept free of dust and dirt.

HOW TO MEASURE CURRENT

The Forestry Service measures the flow of water in mountain streams by determining the gallons of water per minute which pass the measuring station. In electricity the flow is measured by a specified number of electrons passing any given point per second. This specified amount is called a COULOMB. This specified amount or COULOMB flowing per second is equal to ONE AMPERE. A meter used to measure CURRENT must be so connected that ALL of the current flowing in the circuit will flow through the meter.

This is expressed as "a METER CONNECTED IN SERIES in the circuit." This is shown in Fig. 3-8. The meter used is an AMMETER. Note the SYMBOL used to represent this meter.

PREFIXES USED

Prefixes are used the same as with voltage. Most currents which you will measure will be in milliamperes, (one thousandth of an ampere) and amperes. Study the table, Fig. 3-9.

1 AMPERE	= 1 AMPERE
1 MILLIAMPERE	= 1/1000 AMPERE OR .001 AMPERE
1 MICROAMPERE	= 1/1000000 OR .000001 AMPERE

Fig. 3-9. Prefixes used in measuring current.

The terms kilo and mega are seldom used with amperes.

HOW TO READ AN AMMETER

Like the voltmeter, the AMMETER has two test leads. The BLACK or NEGATIVE lead is connected to the negative side of the circuit, and the RED or POSITIVE lead to the positive side of the circuit. NOTE: IN ORDER TO MEASURE CURRENT THE CIRCUIT MUST BE DISCONNECTED AND THE METER CONNECTED IN SERIES. If connected wrong, the meter may be destroyed.

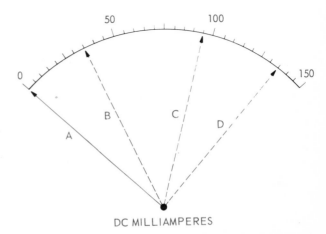

Fig. 3-10. Scale for simple MILLIAMMETER. Pointer in Position A = 0 mA; Position B = 35 mA; Positive C = 95 mA; Position D = 135 mA.

Fig. 3-8. The AMMETER or CURRENT METER MUST BE CONNECTED IN SERIES IN THE CIRCUIT.

Now look at the simple Milliammeter Scale shown in Fig. 3-10. The scale reads 0, 50, 100 and 150. The greatest current you can measure with this scale is 150 milliamperes. Between the major scale divisions are unnumbered marks representing 25, 75 and 125 milliamperes. These divisions are further divided in smaller marks representing 5 mA (note abbreviation) each. The pointer in Fig. 3-10 is shown in four positions. Study the readings.

CURRENT AND AMMETERS

To measure current flowing in a circuit, the conductor must be disconnected and the ammeter reconnected in SERIES with the circuit to be measured. The ammeter has very low internal resistance and should not disturb the circuit resistance.

Fig. 3-11. The *SCALE* and *RANGE SWITCH* for a multirange ammeter.

USING A MULTIMETER

Frequently currents smaller and larger than 150 mA must be measured. For these you will need a multirange ammeter. In Fig. 3-11 the scale of a typical meter is sketched. It is easy to read since the scale reads only 0 to 10. The value of the scale

will be determined by the RANGE SWITCH. Ranges on this meter are 100 μA (microamperes), 10 mA, 100 mA, 1000 mA and 10 A. Study the table, Fig. 3-12.

RANGE SWITCH POSITION	SCALE READS	EACH SMALL MARK EQUALS
100 μA	0 to 100 μA	2 μA
10 mA	0 to 10 mA	.2 mA
100 mA	0 to 100 mA	2 mA
1000 mA	0 to 1000 mA	20 mA
10 A	0 to 10 A	.2 A

Fig. 3-12. *Range Switch Position and Scale Value.*

The indicating needle is shown in four positions in Fig. 3-11. Read the meter and compare your results to the answers given in the table, Fig. 3-13.

RANGE SWITCH POSITION	CURRENT VALUE AT INDICATOR POSITION			
	A	B	C	D
100 μA	20 μA	46 μA	70 μA	92 μA
10 mA	2 mA	4.6 mA	7 mA	9.2 mA
100 mA	20 mA	46 mA	70 mA	92 mA
1000 mA	200 mA	460 mA	700 mA	920 mA
10 A	2 A	4.6 A	7 A	9.2 A

Fig. 3-13. *Answers to Ammeter Reading Experience.*

RESISTANCE

If you turn the faucet in the kitchen sink just a little bit, a small stream of water will flow. If you turn it more a larger stream of water will flow. We can express this another way. As you turn the faucet ON there is less RESISTANCE to the flow of water and therefore more water will flow. RESISTANCE may be defined as an "opposing or retarding force." In electricity resistance is the opposition to the flow of an electrical current by a material, substance, component or device. In electricity the unit of resistance is called an OHM. The Greek letter omega (Ω) is used as the symbol for ohms. Example: one hundred ohms is written as 100 Ω. Its letter symbol is R.

MATERIALS WHICH CONDUCT ELECTRICITY

Materials have various abilities to conduct electricity. This is in part due to their atomic structure and the availability of "free electrons" in the material for conduction purposes. Silver is one of our best conductors. Copper is also a good conductor and is cheaper than silver. Iron and aluminum are not as good as copper. Materials such as glass and rubber which are not good conductors, are called INSULATORS.

A WIDER ROAD ALLOWS MORE TRAFFIC

The size of the wire conductor used determines the total resistance of the wire. As you know, a wide highway allows more automobiles to travel per hour without a traffic jam than a narrow road. We can say, therefore, that a larger and wider road has LESS RESISTANCE or opposition to the flow of traffic than a narrow road. Wire conductors behave in a similar way. THE LARGER THE WIRE THE LESS RESISTANCE IT HAS TO THE FLOW OF ELECTRONS.

LONG ROAD HAS MORE RESISTANCE

If a particular section of narrow road opposes the flow of traffic a certain amount, a section of narrow road four times that long could be expected to oppose the traffic four times as much. In electrical conductors it is the same story. If a wire 10 ft. long has 1 ohm of resistance, then 100 ft. of the same wire would have ten times as much resistance, or 10 ohms. Longer wires must also be larger wires if they are to carry the required amount of CURRENT.

HEAT

Most materials used as conductors will increase in resistance if allowed to become warm. If wires are enclosed in conduit or pipe where they cannot radiate the heat into space, larger wires of less resistance should be used.

WHAT HAVE WE LEARNED ABOUT RESISTANCE

1. It is measured in ohms.
2. The Greek symbol for ohms is Ω.
3. The letter symbol for resistance is R.
4. Resistance depends upon the material used to make the conductor.
5. Larger conductors have less resistance.
6. Longer wires have more resistance.
7. Heat will increase the resistance of a wire.

HOW TO MEASURE RESISTANCE

Resistance is measured with an OHMMETER. Whenever resistance is placed in a circuit it will be represented by this symbol:

Bearing this in mind see Fig. 3-14. In case 1, a wire is connected directly across a voltage source. The wire is so short it is assumed to have ZERO RESISTANCE. This is called a SHORT CIRCUIT and a maximum current will flow.

Case 2. Here two wires are connected to a voltage source but the wires are not con-

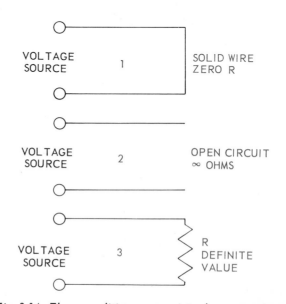

Fig. 3-14. Three conditions can exist when measuring resistance in a circuit.

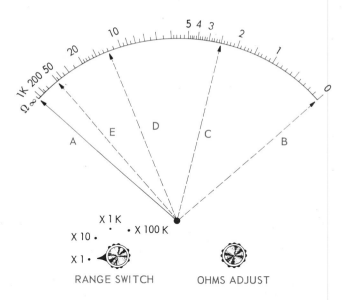

Fig. 3-15. Typical OHMMETER SCALE with RANGE SWITCH and OHM'S ADJUST KNOB.

nected together. This is called an OPEN CIRCUIT. There is NO current path and no current will flow. Its resistance is infinity. The symbol for infinity is ∞. The circuit has R = ∞ ohms.

Case 3. In this circuit a definite value of resistance, represented by the symbol R, has been placed in wires between the voltage terminals. In this case the resistance is not zero, not infinity, but of some value. Current will flow in the circuit depending on the value of the voltage and resistance.

OHMMETER SCALE

The ends of a scale used to measure resistance must be at zero and infinity. All points between these two limits will have

some definite value. In Fig. 3-15 is a typical scale used on the OHMMETER. Also illustrated are a RANGE SWITCH and a knob called OHM'S ADJUST.

Before measuring ohms, the meter must be ADJUSTED TO ZERO. Proceed as follows. Take the two TEST LEADS and touch them together. This is ZERO resistance. Adjust meter with OHM'S ADJUST knob until the meter reads ZERO.

Ohms will be read on the scale directly, but must be multiplied by the setting of the range switch:

SCALE × 1	VALUE AS READ
SCALE × 10	MULTIPLY READING BY 10
SCALE × 1K	MULTIPLY READING BY 1000
SCALE × 100K	MULTIPLY READING BY 100,000

In Fig. 3-15 the indicating needle is shown in five different positions. Read this meter and compare your answers with the table, Fig. 3-16.

OHM'S ADJUST AND AGE

The batteries in the ohmmeter grow weak by age and continued use. Always bring test leads together (zero resistance) and adjust to zero on meter scale, before measuring resistance.

RESISTOR COLOR CODE

In your electronics studies you will be using various kinds of resistors. Many carbon type resistors have bright colored bands.

RANGE SWITCH POSITION	VALUE IN OHMS AT INDICATOR POSITION				
	A	B	C	D	E
× 1	OPEN	SHORT	2.5 Ω	12 Ω	50 Ω
× 10	CIRCUIT	CIRCUIT	25 Ω	120 Ω	500 Ω
× 1K	∞	ZERO	2500 Ω	12 KΩ	50 KΩ
× 100K	OHMS	OHMS	250 KΩ	1.2 MΩ	5 MΩ

NOTE: K means KILO or 1000. M means MEGA or 1,000,000.

Fig. 3-16. Answers to Ohmmeter Reading Experiences.

These bands will tell you the value of the resistors in ohms. The color code, Fig. 3-17, is presented for your study. This should be memorized.

COLOR	NUMERICAL FIGURE	MULTIPLIER	TOLERANCE
BLACK	0	x 1	
BROWN	1	x 10	
RED	2	x 100	
ORANGE	3	x 1000	
YELLOW	4	x 10,000	
GREEN	5	x 100,000	
BLUE	6	x 1 MILLION	
VIOLET	7	x 10 MILLION	
GRAY	8	- - -	
WHITE	9	- - -	
SILVER	-	x .01	±10%
GOLD	-	x .1	±5%
NONE	-	- - -	±20%

Fig. 3-17. Resistor Color Code.

READING RESISTOR COLOR CODE

To identify a resistor from the code, hold the resistor in your hand with the color bands on the LEFT. The first band color is the first number of the value. The second band color is the second number of the value. The third band color tells you to multiply the first two numbers by this factor. In Fig. 3-18, a resistor with bands of BROWN, BLACK, GREEN and SILVER represents 1,000,000 Ω or 1 megohm.

TOLERANCE
NUMBER ZEROS TO ADD
SECOND NUMBER
FIRST NUMBER

1   0   00000   10%

BROWN   BLACK   GREEN   SILVER

Fig. 3-18. Use this example to learn to read the RESISTOR COLOR CODE.

The fourth band, silver, tells you how accurate the resistor may be in order to pass inspection. This resistor is ± 10 percent. It could actually measure as high as 1.1 megohms or as low as 900 KΩ and still be acceptable, (10 percent above to 10 percent below specified value). The more accurate resistors are the more expensive. Most expensive equipment will use ± 5 percent resistors and ± 1 percent resistors. Practice with the color code is the best way to learn it. You will be given many chances to practice.

MULTIMETERS

For convenience, the three meters you have learned to read are available in one case. It is called a VOM (volt-ohm-meter). See Fig. 3-19. The center switch must be used to select WHAT YOU WANT TO MEASURE (VOLTS, AMPERES, or OHMS) and WHAT RANGE YOU WISH TO USE.

Fig. 3-19. A typical Volt-Ohm-Meter, VOM, is a combination meter to measure voltage, amperes and ohms. (Triplett)

STEPS TO UNDERSTANDING ELECTRICITY-ELECTRONICS

1. A VOLT is the unit of measurement of electrical pressure. Its letter symbol is E or V.
2. POTENTIAL DIFFERENCE or ELECTROMOTIVE FORCE is measured in volts.
3. Voltage is always measured between two points in an electrical circuit.
4. The negative or black lead of a voltmeter must always be connected to the point of lowest or most negative potential.
5. The flow of electricity is called CURRENT. Its letter symbol is I. It is measured in amperes and milliamperes.
6. The ammeter must always be connected in SERIES in the circuits so that all current will flow through it.
7. The resistance to the flow of electricity is measured in OHMS. Its letter symbol is R. The Greek letter omega (Ω) means OHMS.
8. A short circuit represents ZERO OHMS.
9. An open circuit represents infinity (∞) ohms.
10. A multimeter measures volts, amperes and ohms in several ranges.

TEST YOUR KNOWLEDGE - Unit 3

1. Select five resistors and record this information:

	COLOR CODE	VALUE IN OHMS	TOLERANCE
R₁	_____	_____ Ω	_____
R₂	_____	_____ Ω	_____
R₃	_____	_____ Ω	_____
R₄	_____	_____ Ω	_____
R₅	_____	_____ Ω	_____

2. Ask your teacher for an ohmmeter. Measure each of your resistors and record their values. Are they within their range of tolerance?

	MEASURED VALUE	ACCEPTABLE WITHIN TOLERANCE
R₁	_____	Yes or No _____
R₂	_____	Yes or No _____
R₃	_____	Yes or No _____
R₄	_____	Yes or No _____
R₅	_____	Yes or No _____

Did you ZERO the ohmmeter before making your ohms measurements?

Fig. 3-20. *Power supply and voltmeter connections for learn-ing to use a voltmeter.*

3. LEARN TO USE A VOLTMETER. Refer to Fig. 3-20. Connect the Voltmeter to a variable voltage supply (0-15 volts). Set voltage control knob to minimum and TURN ON the supply. Adjust voltage to read 1 volt, 3.5 volts, 5 volts, 8 volts, 11 volts, 13.5 volts and 15 volts.

4. Disconnect your voltmeter from the power supply. Turn the control knob to any position. Now reconnect the voltmeter and record the voltage. Disconnect voltmeter again and reset control knob at another po-sition. Measure the voltage. Repeat this ex-perience several times until you can mea-sure and read any voltage.

5. LEARNING TO USE THE AMMETER. In this experience you will need a LOAD which will draw some current from this power source. Select a 100 ohm resistor and connect it in a circuit as in Fig. 3-21.

Fig. 3-21. *Connection of circuit to measure* **CURRENT.**

Start from the negative terminal and trace the path of electron flow through the RESISTOR, through the AMMETER and then to the positive terminal of the power source. Turn on the power source and increase its voltage to maximum value. Note the CURRENT value on the milliammeter.

BUILD A COLOR CODE VISUAL AID

A colorful and useful device for learning the color code is shown in Fig. 3-22. The colored rectangles are plastic and can usually be obtained from the scrap pile of a company that makes neon signs and lights. Use your own creativeness and imagination on the overall design.

OHMMETERS ARE SELF-POWERED

When measuring resistances in a piece of equipment be sure the equipment is not "turned on." Better to pull out the plug from the power source. An ohmmeter has its own battery inside the meter case.

WHAT IS YOUR BODY RESISTANCE?

Hold one probe of the ohmmeter between the fingers of your right hand and the other probe in your left hand. What is your resistance to the flow of current? Now wet your fingers and repeat the measurement. Is there a difference? Is it safe to work around electricity in wet places?

Fig. 3-22. A COLOR CODE DEVICE for shop or laboratory. (Courtesy of Prof. Jason McLoney, San Francisco State College)

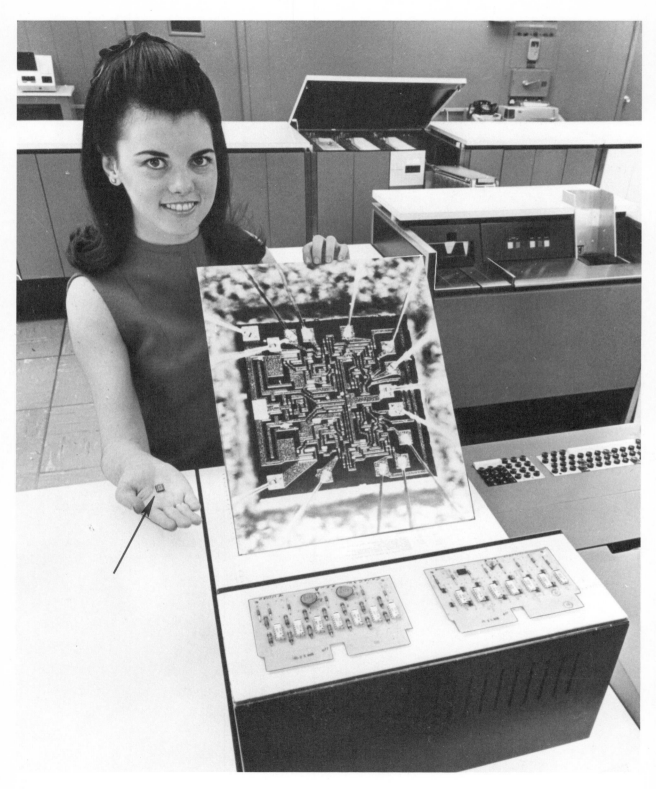

The young lady is holding an integrated circuit in one hand (arrow) and a photographic enlarge-ment of a circuit in the other hand. Circuits like these carry electronic messages inside computers. (Honeywell Electronic Data Processing)

Unit 4

POWER AND WORK

In the previous Unit you learned about the VOLT, the AMPERE, and the OHM. In this Unit, you will learn about the relationship of these terms.

VOLTAGE, CURRENT AND RESISTANCE

The circuit and meters to be used are shown in Fig. 4-1. The schematic diagram appears in Fig. 4-2.

SOLVING PROBLEM 1. How does a change in VOLTS affect an electrical circuit? Turn on the power source and adjust the voltage to 5 volts. What is the current? Now increase the voltage to 10 volts. Read the milliammeter. Increase the voltage to 15 volts

Fig. 4-2. Schematic diagram for the circuit in Fig. 4-1.

and read the milliammeter. What are your conclusions? Solution: As you might expect, as the pressure is INCREASED, the flow is

Fig. 4-1. Circuit set up to determine the relationship between VOLTS, and AMPERES.

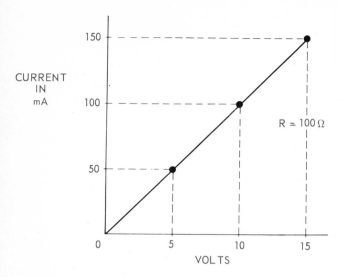

Fig. 4-3. A graph to show the relationship between voltage and current. Resistance is 100 ohms.

rent at ANY voltage between 0 and 15 volts. There is a DIRECT RELATIONSHIP between voltage and current.

SOLVING PROBLEM 2. What effect does resistance have on a circuit? Leave the fixed $100\,\Omega$ resistance in your circuit and add a variable $1000\,\Omega$ resistor in series. The re-drawn circuit is shown in Fig. 4-4. Fig. 4-5 shows a schematic for the circuit. Set the voltage at 15 volts and observe the current. Now turn the knob of the variable resistor TO INCREASE ITS RESISTANCE. What happens to the current? Turn the variable resistance in the opposite direction TO DECREASE ITS RESISTANCE. What happens to the current?

SOLUTION: Here is proof to another basic fact about electrical circuits. If you DE-CREASE THE RESISTANCE THERE IS IN-CREASED CURRENT FLOW. If resistance is INCREASED, the CURRENT must DECREASE.

INCREASED. In electricity, if the voltage is increased, the current will increase. Also, if voltage is decreased, the current will decrease.

In Fig. 4-3 a GRAPH shows you how the engineer expresses this relationship. By using this type of graph, you can find the cur-

This relationship may be expressed in an upside down fashion. CURRENT IS IN-VERSELY related to the RESISTANCE in a circuit:

Fig. 4-4. Circuit which may be used to prove the relationship between current and resistance in a circuit.

R increases, I decreases

or

R decreases, I increases

These voltage, current and resistance relationships are expressed by the well known OHM'S LAW. This law states that:

$$I = \frac{E}{R} \quad \text{or} \quad R = \frac{E}{I} \quad \text{or} \quad E = I \times R$$

when I = amperes, R = ohms, and E = volts

You can see that if any two values are known, the third and unknown can be found by simple mathematics.

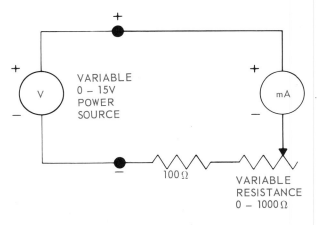

Fig. 4-5. The schematic for circuit Fig. 4-4.

WORK

We learn about WORK at any early age. Mow the lawn. Run errands. Carry some wood. Study. All these jobs are work and require physical effort and the use of ENERGY. Lift a ten pound stone to a height of one foot. You have done ten foot-pounds of work. When a force moves through a distance, WORK is done. The definition in physics will say that:

work = force x distance

Look at a lawn mower. Let's assume it requires 10 pounds of force to push it through the grass. If your lawn is fifty feet wide, then you would do 500 foot-pounds of work by pushing it once across the lawn. It takes a lot of work to mow a lawn. Energy is used up.

POWER

Power is defined as the RATE OF DOING WORK. How long does it take you to push the lawn mower across the lawn just once? One minute? Ten minutes? Thirty seconds? In the first example, you would do 500 ft. lbs. of work per minute. In the second, 500 ft. lbs. of work per 10 minutes and in the third example, 500 ft. lbs. of work per 30 seconds.

James Watt, inventor of the steam engine, looked around for a way of determining the power of his newly invented engine. In these days, the farm horse did most of the work, so Watt compared his engine to the rate of work of the horse. As a result we have the familiar term "horsepower." A one horsepower rating states that 33,000 foot-pounds of work are done in one minute or 550 foot-pounds per second.

WATTS

In electricity, we can now state that VOLTAGE is a force and the movement of a quantity of electrons per second is an AMPERE. We arrive at the conclusion that VOLTS times AMPERES equals POWER. So that James Watt will not be forgotten, the unit of electrical power is named a WATT. To find the rate of the use of energy or power of a circuit, it is:

VOLTS times AMPERES equals WATTS
or
E x I = P in watts

WATT-HOURS

We have been reading about the rate of using electrical energy, but another point must be considered. How long did you work at that rate? If you worked for one hour, you then could determine the total energy used. If an electrical device such as a lamp or a motor is using power at the rate of 100 watts, at the end of one hour, it would have used

100 WATT-HOURS of electrical energy. The power company sells power to your home at an established rate per KILOWATT-HOUR (KWH). The meter measures the amount of power used. At the end of the month a bill is sent.

EXAMPLE 1. Let's assume that electricity costs five cents per KWH in your locality. At night your family burns ten one hundred watt lights for five hours. How much does it cost?

$$100 \text{ watts} \times 10 \times 5 = 5000 \text{ watt-hours}$$
$$5000 \text{ watt-hour} = 5 \text{ KWH}$$
$$5 \text{ KWH} \times 5 \text{ cents} = 25 \text{ cents}$$

EXAMPLE 2. Your home has a voltage of 110 volts. A toaster draws 5 amperes of current and you toast bread for 30 minutes. How much does it cost if energy costs 5 cents per KWH or .005 cents per watt-hour?

$$P = I \times E \quad \text{so}$$
$$WH = 550 \times .5 \text{ hrs.} = 275 \text{ watt-hours}$$
$$275 \text{ watt-hours} \times .005 \text{ cents} = 1.375 \text{ cents}$$

EXAMPLE 3. The water heater in your home has a 2000 watt heating element. At a voltage of 100 volts how much current does it use?

$$I \times E = P \qquad I \times 100 = 2000 \text{ W}$$
$$I = 20 \text{ A}$$

In the table, Fig. 4-6, are listed some common appliances used in the home and their consumption of power in watts.

Power companies continue to supply an almost unlimited quantity of electrical energy. Some is generated by hydroelectric plants which convert the energy of falling water to electric power. In many locations steam produced by coal, gas or oil-fueled boilers turns the generators. The use of atomic energy is gaining in popularity and may become our major source of power in the future.

READING YOUR ELECTRIC METER

The arrival of the "meter reader" from the power company is generally announced by the dogs in the neighborhood. Sometimes over-protective dogs make the life of a meter man a hazardous occupation.

Recent electronic equipment now being tried, automatically reads meters by electronic circuitry and telephones this information to the business office computer of the power company. The meter reader may become a "lost" occupation.

Your house meter will have four dials with numbers on each reading 1 to 10. Refer to Fig. 4-7. The first dial on the right reads in kilowatt-hours. The second dial reads in ten kilowatt-hours. The third dial reads in 100 kilowatt-hours. The fourth dial reads in 1000 kilowatt-hours. NOTE: When reading any of the four dials always use the number which the pointer has just passed. If the pointer is between 6 and 7, use 6. If the pointer seems to be on a number refer to the dial on its right to determine if the pointer is exactly on the number or a little before or after the number. The examples in Fig. 4-7 will help you to understand.

Let's assume that one month ago your meter read 3152, and today it reads 4827. The power consumed for one month would be, 4827 - 3152 or 1675 KWH. At a typical cost of 5 cents per KWH, the electric bill would be, 1675 x .05 = $83.75. However the rate usually changes to a lower price if you use a large quantity of power, such as:

First 50 KWH at .05
Next 100 KWH at .03
All above 150 KWH at .02

DEVICE	POWER IN WATTS
Lamp Bulbs	3–300
Radio	50–150
Television	200–500
Refrigerator	250–300
Electric Range	8000–16000
Electric Iron, Toaster	500–1000
Motor – 1/4 hp	250–300

Fig. 4-6. Table which shows typical home appliances and power ratings.

EXAMPLE A
METER READS 3152 KWH

EXAMPLE B
METER READS 4827 KWH

*Fig. 4-7. Examples of Meter Reading. **EXAMPLE A**: Meter Reads 3152 KWH. **EXAMPLE B**: Meter Reads 4827. Note Example B, that dial 2 (second from right) reads 2, not 3. It will not be 3 until dial 1 reaches 0. The same with dial 3. It will not be 9 until dial 2 reaches 0.*

Using these rates the above bill would be:

50 KWH at .05	=	2.50
100 KWH at .03	=	3.00
1525 KWH at .02	=	30.50
Total cost		$36.00

FORWARD STEPS IN UNDERSTANDING ELECTRICITY-ELECTRONICS

1. As voltage of a circuit increases, current will also increase.
2. If resistance in a circuit is increased, the current will decrease.
3. Voltage and Current are directly related.
4. Current and Resistance are inversely related.
5. Power is the rate of doing work.
6. Electrical power is measured in Watts.
7. Voltage multiplied by Current equals Power.
8. Energy consumed is measured in watt-hours.
9. A kilowatt-hour equals one thousand watt-hours.
10. Electricity is purchased by the KWH unit of measure.

TEST YOUR KNOWLEDGE - UNIT 4

1. Read your electric meter today and again in one week. Inquire from your power company the cost of electric power per KWH. Figure the cost of the power used for the week.
2. A circuit has an applied voltage of 100 volts and a circuit current of 2 amperes. What is its power?
3. If your color TV uses 500 watts of power, how much does it cost to watch a two hour program at 5 cents per KWH?
4. The resistance of an electric roaster is 20 ohms. With a home voltage of 120 volts, how much power does the roaster use?
5. For Christmas decoration of your home, ten strings of colored lamps are used which use 50 watts of power each. How much current will be required? Use voltage of 120v.
6. Explain what will happen if a short circuit should happen in a home appliance.
7. Connect the circuit with a variable resistor as in Fig. 4-4. Adjust voltage to maximum of 15 volts. Adjust resistance for a 100 mA current.

    Turn voltage to zero. Record current.
    Repeat in steps for 1, 2, 5, 10 and 15 volts. Record current after each step.
    Draw a graph of your results.

Unit 5

SOURCES OF ELECTRICAL ENERGY

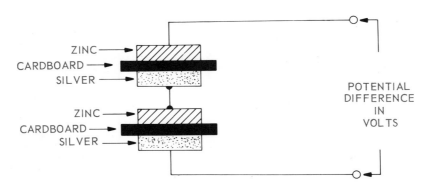

Fig. 5-1. The original BATTERY consisted of zinc and silver pieces of metal separated by cardboard soaked in salt solution.

Down through the ages, men of science have experimented with methods of creating a Potential Difference which will cause electric current to flow. The engineer has applied his scientific "know-how" to make the electric current perform useful work.

CHEMISTRY AND ELECTRICITY

The first practical electric BATTERY was demonstrated by the Italian scientist, Allessandro Volta, in about 1800. The battery consisted of pieces of zinc and silver separated by cardboard soaked in salt solution. See Fig. 5-1. Each pair of zinc and silver pieces is considered a CELL. Cells connected together as shown make a BATTERY. The conclusion drawn from Volta's work is: If two dissimilar metals called ELECTRODES are placed in a solution, called the ELECTROLYTE, which chemically reacts with the metals, a POTENTIAL DIFFERENCE or VOLTAGE will be produced between the two metals. The chemical reaction causes one electrode to LOSE ELEC-

TRONS and therefore becomes POSITIVE. The other electrode GAINS ELECTRONS and becomes NEGATIVE. In the CELL, CHEMICAL ENERGY is converted to ELECTRICAL ENERGY.

EXPERIMENT 1. Secure a penny, a dime and a small piece of blotting paper. Soak the blotting paper in salt water. Clamp the parts together with a clothespin as in Fig. 5-2. Measure the voltage with your meter. Is this an efficient cell? Is the dime or the penny negative?

Fig. 5-2. A simple cell made with a penny and a dime.

EXPERIMENT 2. A lemon CELL. Secure a lemon or a grapefruit, a strip of zinc and a strip of copper. Stick the metal electrodes into the lemon as shown in Fig. 5-3. What is the voltage developed? Which metal is negative? Which is positive? Try inserting the metal strips closer together. Does the voltage increase or decrease?

Fig. 5-4. A zinc-carbon cell will produce a voltage by chemical action.

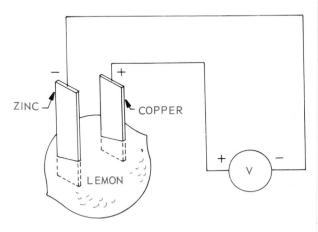

Fig. 5-3. A lemon cell made with copper and zinc will produce a voltage.

EXPERIMENT 3. Secure a small glass jar and fasten a zinc electrode and a carbon electrode in an upright position. See Fig. 5-4. Fill the jar three-quarters full of a strong salt solution. Measure the voltage. Which electrode is positive or negative? Does this cell produce a larger voltage than either the penny-dime or lemon cell? While the voltmeter is connected to the cell, connect a 100 ohm resistor to the cell. The voltage

of the cell will drop. With a load attached and drawing current, this inefficient cell cannot hold its voltage.

COMMERCIAL CELLS

The flashlight CELL and the so-called DRY CELL are examples of cells which you may purchase from a hardware or electronics store. Typical cells appear as in Fig. 5-5. You may compare the commercial cell with your experimental cell. Study the cutaway view in Fig. 5-6. Note that it is assembled in a ZINC can (zinc electrode). In the center

Fig. 5-5. A wide selection of mercury cells provides a full selection for industrial, instrument and electronic applications. (Burgess Battery Div., Clevite Corp.)

POSITIVE TERMINAL
BINDING POST

NEGATIVE TERMINAL
BINDING POST

COVER
PLASTIC COATED
INSULATION BOARD

INNER SEAL
ASPHALT

EXPANSION
CHAMBERS

SEAL SUPPORT
WASHER

CARBON
ELECTRODE

PLASTIC COATED
PULPBOARD
SEPARATOR

DEPOLARIZING MIX

ZINC CAN
OUTSIDE SURFACE
ASPHALT COATED

CHIPBOARD
JACKET

Fig. 5-6. A cutaway view of a No. 6 Dry Cell.

is the carbon rod electrode. The electrolyte is made of absorbent material with a mixture of sal ammoniac and zinc chloride.

A serious fault developed with early cells. As a result of the chemical action, bubbles of hydrogen gas formed around the carbon electrode. This was like an insulating "blanket" and caused the cell action to stop. This action is called POLARIZATION. To overcome this defect a depolarizing agent is added to the electrolyte which combines with the hydrogen to form water.

Do not leave dead cells in your flashlight. Dead cells sometimes expand and corrode as a result of chemical action and the formation of water.

Cells used in flashlights are called PRI-

MARY CELLS because they generally cannot be recharged. The chemical action is not reversible. However, a cell charger now on the market, can revitalize a cell and increase its useful life.

ACIDS EAT AND BURN

Many batteries, like the automotive battery, will use an acid electrolyte. Drops of this acid can burn your hands and eat holes in your clothing. You must use great care when working around them. Acid burns can be neutralized with bicarbonate of soda.

THE AUTOMOTIVE BATTERY

Examples of cells which can be recharged many times are the lead-acid cells used in the automotive storage battery. These cells

Fig. 5-7. A cutaway view of a typical automotive battery.
(ESB Brands, Inc.)

are SECONDARY CELLS. A complete battery is illustrated in Fig. 5-7. Here you can see the electrodes or plates and the construction of the battery. The negative plates are made of spongy lead. The positive plates are lead peroxide. The electrolyte is sulfuric acid and water.

During discharge the acid "goes into the plates" and the plates chemically change to lead sulfate. The electrolyte becomes nearer to pure water.

A setup for recharging a battery is shown in Fig. 5-8. An external power source is connected so a current is forced through the battery in the opposite direction. This reverses the chemical action. The plates return to their original state. The acid leaves the plates and makes the electrolyte richer in acid. A convenient way to check the "state

of charge" of the battery is with a HYDROMETER. In a hydrometer a calibrated floating glass tube sinks less in a liquid heavy with acid than in light liquid. The next time you visit a service station or your school automotive shop, ask the attendant or teacher to show you a HYDROMETER. A similar hydrometer is also used to check the amount of antifreeze in the radiator of your car.

A RECHARGEABLE DRY CELL

In the last few years many common electric appliances have appeared on the market which are self-powered. These contain their own batteries, which can be recharged by plugging into the home power source. Examples of these include flashlights, electric razors, electric toothbrushes, electric carving knives, electric drills and many others. The cells usually are nickel-cadmium cells

Fig. 5-8. A battery charger connected to battery to show reverse CURRENT which reverses chemical action.

with sintered powdered nickel plates formed on a nickel wire screen. The positive plate is treated with nickel salt solutions. The negative plate uses cadmium salts. The electrolyte is potassium hydroxide. An explanation of the chemical action is beyond the scope of this text.

USING CELLS IN SERIES

In most applications a single cell will not produce high enough voltage to do the re-

quired job. Secure a flashlight cell and measure its voltage. It will be about 1.5 volts. When a higher voltage is needed CELLS are connected in SERIES. Two or more cells connected together make a BATTERY.

The series method is shown in Fig. 5-9. The schematic is also shown so that you will recognize this easy way to show a circuit. Note that now the total output voltage is equal to the voltage of one cell (1.5 volts)

Fig. 5-9. Cells connected in SERIES, NEGATIVE to POSITIVE. Total voltage equals 6v.

Fig. 5-10. A flashlight using three D cells in SERIES. Total voltage about 4.5v.

times the number of cells (4) or 1.5v x 4 = 6 volts. To connect cells in SERIES, the negative of one cell is connected to the positive of the next and so on. You might say the cells are connected in line or end to end. A common example of this type of battery is in a 3 cell flashlight. See Fig. 5-10.

Frequently cells are connected to produce a current over a long period of time, when an increase in voltage is not needed. This is the PARALLEL or SIDE BY SIDE connection. In Fig. 5-11 the four cells are connected in this manner. The total voltage remains only 1.5 volts. The capacity to supply current has increased about four times. Each cell contributes its part to the total current.

INTERESTING EXPERIENCES WITH CELLS

Secure four flashlight cells from your school's supply. Measure the voltage of each cell.

1. Connect two cells in series.
   Measure the voltage.
2. Connect three cells in series.
   Measure the voltage.
3. Connect four cells in series.
   Measure the voltage.
4. Connect two cells in parallel.
   Measure the voltage.
5. Connect three cells in parallel.
   Measure the voltage.
6. Connect four cells in parallel.
   Measure the voltage.

Look over an automotive storage battery. If of the 12-volt type, it will have six cells of 2 volts each in series.

ELECTRICAL ENERGY FROM LIGHT ENERGY

One of the more exciting experiences you will have in electronics is the conversion of light to electricity. A device used for this purpose is called a SOLAR CELL. When light

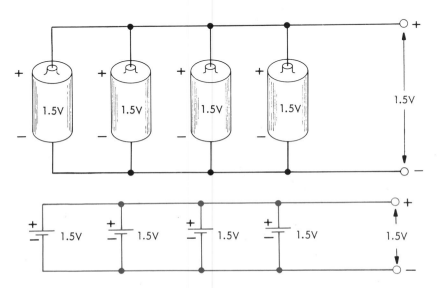

Fig. 5-11. Cells connected in PARALLEL. Terminal voltage remains the same as the voltage of one cell.

Fig. 5-12. *The construction of a photovoltaic cell or* **SOLAR CELL.**

strikes the junction between certain semi-conductor materials, a voltage is produced. A sketch illustrating this phenomenon (fact of scientific interest) appears in Fig. 5-12.

A cell similar to this is used in camera light meters. The brighter the light - the higher the voltage developed. A meter is used to indicate the brightness.

BUILDING SIMPLE LIGHT METER

To build a simple light meter, secure a B2M Solar Cell and connect it to your volt-meter. Follow the diagram, Fig. 5-13. Then

Fig. 5-13. *A simple circuit to measure light intensity.*

direct a flashlight on the solar cell and observe the movement of the meter.

An improved light meter, as shown in Fig. 5-14, uses a battery and a transistor to increase its sensitivity to light change. You will study transistors later in this text.

PHOTOCONDUCTIVE CELLS

Another component which takes advantage of light intensity is a cell which CHANGES

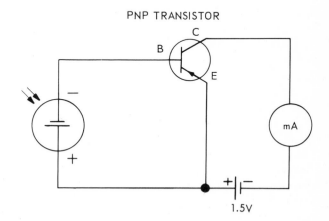

Fig. 5-14. *An improved light meter using a* **TRANSISTOR** *to increase its sensitivity.*

ITS RESISTANCE when exposed to light. A sketch of such a device and its symbol will be found in Fig. 5-15. A photoconductive material (cadmium sulfide) is supported in the center of a wire mesh. When light shines on

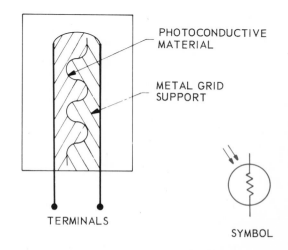

Fig. 5-15. *The construction and electronic symbol for a Cadmium-Sulfide photoconductive cell.*

this material, it will increase its conductivity or decrease its resistance. Notice that this cell controls the current in a circuit and can be used directly to operate meters and relays.

CHANGING RESISTANCE BY LIGHT

In the circuit, Fig. 5-16, the photoconductive cell is connected in series with the current meter and a power source. As the intensity of the light increases, the current also increases. Such devices are widely used

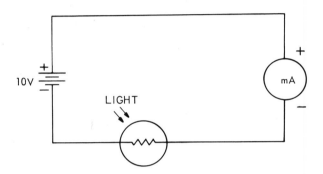

Fig. 5-16. A light sensitive circuit using the photoconductive cell. The brighter the light, the higher the current.

in industry for counting and sorting, fire and burglar alarms, automatic door openers and safety warning systems and for reading information into computers. Some interesting circuits will be described in later Units.

ELECTRICITY FROM HEAT

An unusual effect takes place when two dissimilar metals are joined together and heated. A slight potential difference is produced. This device is known as a THERMO-COUPLE. An experiment which you may perform is illustrated in Fig. 5-17. The dissimilar wires are iron and copper. The commercial type of thermocouple may use wires of copper-constantan, iron-constantan or platinum-platinum rhodium. Constantan is an alloy of about 40 percent nickel and 60 percent copper. The thermocouple is used industrially for temperature control and monitoring (observing and checking). If a higher output voltage is required, several thermocouples may be connected together to form a THERMOPILE.

Fig. 5-17. A voltage is produced by heating the THERMOCOUPLE.

ELECTRICITY FROM MECHANICAL PRESSURE

A very interesting phenomenon occurs when you bend or apply pressure to certain kinds of crystals such as quartz, tourmaline and certain ceramic materials. Actually the distortion causes the electrons to move out

Fig. 5-18. When a crystal is bent or flexed, the electrons are distorted out of their normal positions.

of their usual orbits and a small voltage is produced between the sides or surfaces of the crystal. When the force is removed, the electrons return to their normal positions. See Fig. 5-18. A laboratory apparatus for demonstrating this effect, called PIEZO-ELECTRIC EFFECT, is shown in Fig. 5-19. Applications of this unusual effect are found in crystal microphones, crystal cartridges for record players and many other industrial circuits. Microphones using a ceramic element appear in Fig. 5-20.

ELECTRICITY FROM MOTION AND MAGNETISM

The major source of electrical power for home and industry is found in the reaction between a moving conductor and a magnetic

Fig. 5-19. A laboratory demonstrator for piezoelectricity. The crystal is flexed by moving the lever. The generated electricity will light the lamp. (Central Scientific Co.)

field. It is the GENERATOR. In Unit 8 we will study this most important principle of producing electrical energy.

FORWARD STEPS IN UNDERSTANDING ELECTRICITY - ELECTRONICS

1. Electricity is produced by converting a form of energy to electrical energy.
2. Chemical energy may be converted to electrical energy.
3. Light energy may be converted to electrical energy.
4. Heat energy may be converted to electrical energy.
5. Mechanical energy may be converted to electrical energy.
6. Electrical energy may be produced by movement of a conductor in a magnetic field.
7. Elements in a cell are called electrodes and electrolyte.
8. Cells in series produce a higher voltage.
9. Cells in parallel produce a greater capacity to supply current of the same voltage as each individual cell.
10. A primary cell generally cannot be recharged.

11. The lead-acid cell can be recharged.
12. A battery is two or more cells connected together.
13. Light energy will change the resistance of certain kinds of material.
14. Electrical energy may be produced from heat energy by using a thermocouple.
15. Piezoelectricity is produced by pressure or by flexing a crystal.

Fig. 5-20. Typical ceramic type microphones used with Citizens Band, Amateur, Fixed Station and Mobile radio transmitters. (Shure Bros., Inc.)

TEST YOUR KNOWLEDGE - UNIT 5

1. Six volts are required to operate a small lamp. You have several flashlight cells on hand. Draw the circuit and connect the cells to produce the required brightness.
2. Connect two D cells in series to the 6 volt lamp. Does the lamp glow at full brightness? Explain.
3. Using the ohmmeter, what is the resistance of the 6 volt lamp?
4. Did the lamp glow when you connected the ohmmeter to it? Why?
5. Design a circuit which would electronically tell you if it was day or night. Could you use the same circuit to tell if it is a sunny or cloudy day?
6. Design a circuit using a photoconductive cell by which you can tell if it is night or day.

Unit 6

MAN CONTROLS ELECTRICITY

When we have a potential difference or voltage, we have a source of electrical energy. In early experiences, you learned that it is necessary to connect this voltage to some device or load to produce work or some desirable effect. The connections are made by wires or CONDUCTORS. A conductor is a low resistance path or road for electron current to flow.

It might be well at this time to understand exactly how current flows in a wire. It does not run through wire like water runs through a pipe.

Fig. 6-1 is an enlarged view of a piece of copper wire. Only a few atoms of copper are shown in the drawing. In a given wire there are billions of such atoms. As a negative potential is applied to one end of the wire, the force or energy is transferred from electron to electron to the positive end of the wire or source. At this positive end electrons are attracted off the wire. These facts you should remember:

1. At no time does the conductor store electrons. It gives up one electron at the positive end for each electron forced on at the negative end.

2. The ability of a conductor to conduct

SERIES OF BALLS IN CONTACT WITH EACH OTHER

Fig. 6-2. An experiment to demonstrate that a force applied to one end of a string of balls is transferred through the balls to the opposite end.

depends on the number of free electrons available for conduction.

3. The actual drift of electrons through the wire is relatively slow.

4. The transfer of electrical energy approaches the SPEED OF LIGHT which has been measured as 186,000 miles per second or 300,000,000 meters per second.

In the physics laboratory the transfer of energy may be demonstrated by a line of balls in contact with each other. Refer to Fig. 6-2. As the first ball rolls down and hits the next ball, the energy is transferred through the balls and the last or eighth ball is forced away from the string.

Fig. 6-1. DRIFT of free electrons through a conductor is a current.

ELECTRON FROM NEGATIVE   ELECTRON TO POSITIVE

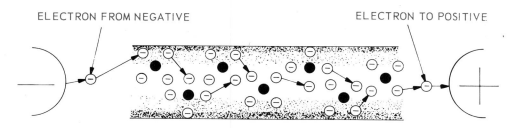

GAGE NO.	DIAM. MILS.	OHMS PER 1,000 FT. OF COPPER WIRE AT 25°C.	GAGE NO.	DIAM. MILS.	OHMS PER 1,000 FT. OF COPPER WIRE AT 25°C.
1	289.3	0.1264	21	28.46	13.05
2	257.6	0.1593	22	25.35	16.46
3	229.4	0.2009	23	25.57	20.76
4	204.3	0.2533	24	20.10	26.17
5	181.9	0.3195	25	17.90	33.00
6	162.0	0.4028	26	15.94	41.62
7	144.3	0.5080	27	14.20	52.48
8	128.5	0.6405	28	12.64	66.17
9	114.4	0.8077	29	11.26	83.44
10	101.9	1.018	30	10.03	105.2
11	90.74	1.284	31	8.928	132.7
12	80.81	1.619	32	7.950	167.3
13	71.96	2.042	33	7.080	211.0
14	64.08	2.575	34	6.305	266.0
15	57.07	3.247	35	5.615	335.0
16	50.82	4.094	36	5.000	423.0
17	45.26	5.163	37	4.453	533.4
18	40.30	6.510	38	3.965	672.6
19	35.89	8.210	39	3.531	848.1
20	31.96	10.35	40	3.145	1069.

Fig. 6-3. Wire Gage and Resistance chart.

SIZES AND TYPES OF CONDUCTORS

Hundreds of types and sizes of wire and cables are required to meet the needs of the electronics industry. You have already learned that the size of wire required is determined by the amount of current it must conduct. A small wire has more resistance than a large wire. A study of the table, Fig. 6-3, will show wire sizes are identified by gage number. The larger the number, the smaller the wire. Also note the resistance per 1000 ft. of each size. Fig. 6-4 shows some common types of wire which you will use in this course.

ELECTRIC CIRCUIT

Electric and electronic circuits may be divided into four parts. Each part is identified by its use. An awareness of these parts of the total system will help you understand how they work and how to service the equipment.
1. The POWER SOURCE.
2. The CONDUCTORS or CURRENT PATHS.
3. The CONTROL.
4. The LOAD or WORK to be done.

In Fig. 6-5 each part is identified in a simple circuit. Fig. 6-6 shows the schematic diagram of the same circuit.

ENAMEL INSULATION
SOLID COPPER WIRE

RUBBER OR PLASTIC INSULATION
HOOKUP WIRE
STRANDED OR SOLID COPPER WIRE

COAXIAL CABLE
RUBBER INSULATION
BRAIDED OUTER CONDUCTOR
PLASTIC INSULATION
STRANDED INNER CONDUCTOR

Fig. 6-4. Three common types of conductors.

Fig. 6-5. The basic electrical circuit with Power Source, Conductors, Control and Load.

Fig. 6-6. The schematic diagram of the basic electrical circuit.

The parts of the basic circuit may take many forms. The SOURCE of ENERGY may be any one of the five sources of energy. The LOAD may be a lamp, a motor, a heater, a radio, a TV or any one of hundreds of other devices. The CONTROL may be a simple "on and off" switch or any number of devices which would limit the current or the voltage of the circuit.

SERIES CIRCUIT

In a SERIES circuit there is only one path for the current to flow and the same current flows through each and every component in the circuit. In Fig. 6-7 a 6 volt power source is connected to:

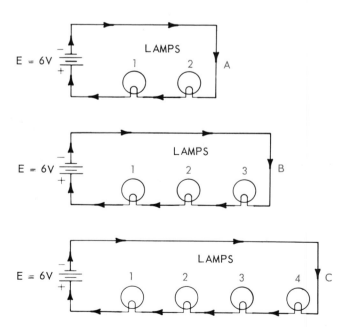

Fig. 6-7. Two, three and four lamps connected in SERIES.

A. Two lamps in SERIES.
B. Three lamps in SERIES.
C. Four lamps in SERIES.
The path of current flow is marked by arrows in each circuit.

EXPERIENCES WITH SERIES CIRCUITS

1. Connect the circuit with a single lamp, as in Fig. 6-8. Observe how brightly the lamp glows.
2. Connect the milliammeter in series with the circuit and measure the current required to light the lamp. The circuit in Fig. 6-8 includes a voltmeter, milliammeter and a single-pole switch.

Fig. 6-8. The circuit used in the text to measure current and voltage in a simple one lamp circuit.

A commercial laboratory setup for this experience is shown in Fig. 6-9.
3. Turn off the power and connect a second lamp in series with Lamp 1. Turn on the power and adjust voltage to six volts. Do the two lights glow as brightly as only one did in your previous experience? Why?
4. What is the current measurement? Why is the current only half as much as in Experience 2? The RESISTANCE of the circuit must be two times greater and the current therefore is half as much.
5. Disconnect the voltmeter from across the power source and reconnect it across Lamp 1 as shown in Fig. 6-10. The voltage should read about 3 volts. Connect the voltmeter across Lamp 2. It also should read about 3 volts.

CONCLUSIONS

1. One lamp required 6 volts and a current of 100 mA to glow brightly.
2. When two lamps were connected in

Fig. 6-9. In this laboratory setup two lamps are connected in series. A meter measures the source voltage. (Lab—Volt, Buck Engineering Co.)

SERIES, the lamps glowed only half as brightly. The voltage across each lamp was only 3 volts and the circuit current was half as much.

3. By adding a second lamp in SERIES the circuit resistance must have doubled. The total resistance of the circuit must be:

R Lamp 1 + R Lamp 2 = TOTAL R

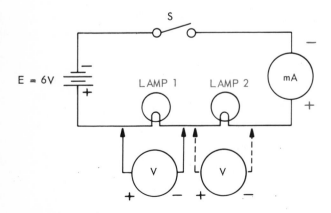

Fig. 6-10. This circuit illustrates method of measuring voltage drops across the lamps.

4. The voltage across Lamp 1 plus the voltage across Lamp 2 is equal to the source voltage or,

3v + 3v = 6v (source voltage)

5. The voltage loss across either lamp is called the "voltage drop" and the sum of the voltage drops equals the source voltage.

6. The voltage drop across a component can always be found by the formula of current times resistance,

I Circuit x R Component = Voltage Drop

$$I \times R = E$$

The voltage drop is frequently called the IR drop.

Repeat the Experiences with SERIES CIRCUITS using 3 lamps and 4 lamps until you are satisfied that you completely understand these conclusions. These experiences are among the most important lessons in your study of electricity.

PARALLEL CIRCUITS

When components are connected, side by side, so that there is more than one path for current flow, the components are connected in PARALLEL. In Fig. 6-11 a six volt power source is connected to:

A. Lamp 1 only.
B. Lamp 1 in parallel with Lamp 2.
C. Lamp 1, Lamp 2, Lamp 3 in parallel.
D. Lamp 1, Lamp 2, Lamp 3 and Lamp 4 in parallel.

The paths for the current flow for the lamps are indicated by arrows.

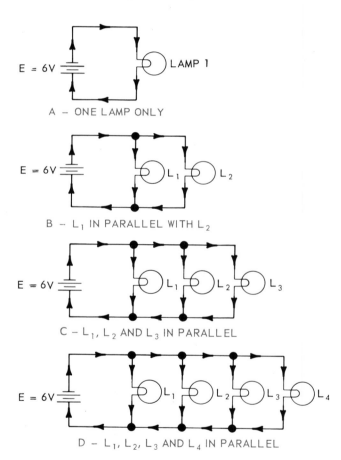

A – ONE LAMP ONLY

B – L_1 IN PARALLEL WITH L_2

C – L_1, L_2 AND L_3 IN PARALLEL

D – L_1, L_2, L_3 AND L_4 IN PARALLEL

Fig. 6-11. One, two, three and four lamps in PARALLEL.

EXPERIENCES WITH PARALLEL CIRCUITS

1. Connect the circuit with power source, milliammeter and voltmeter and one lamp as displayed in schematic drawing Fig. 6-12. Does the lamp glow brightly? What is the

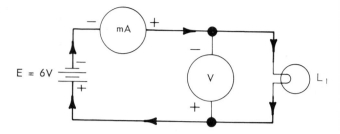

Fig. 6-12. A single lamp connected to voltage source with voltmeter measuring voltage across lamp and current meter to measure TOTAL CURRENT drawn from source.

voltage across the lamp? How much current does this one lamp use?

2. In Fig. 6-13 Lamp 2 is connected in parallel with Lamp 1. Does Lamp 2 glow as brightly as Lamp 1? What is the voltage across L_1? What is the voltage across L_2? Are these voltages the same? But look at the MILLIAMMETER! The current reads twice as much as the single lamp circuit.

Fig. 6-13. Two lamps connected in parallel. The voltage across either lamp is the same as the source. The total current is equal to the SUM of current I_1 and I_2.

CONCLUSION

The total current drawn from the source must equal the current of L_1 plus the current of L_2.

3. Add to your circuit the third Lamp, L_3, according to the schematic in Fig. 6-14.

Fig. 6-14. Three lamps connected in parallel. The voltage across each lamp is the same as the source voltage. The total current, I_T, is equal to the SUM of $I_1 + I_2 + I_3$.

Do L_1, L_2 and L_3 burn with equal brightness? Are the voltages across L_1, L_2 and L_3 the same? But look at the current! It is now three times greater than when L_1 was connected alone.

CONCLUSION

When lamps are connected in parallel, they all glow with equal brightness. Each lamp has the source voltage connected to it. Each lamp added in parallel INCREASES the total circuit current drawn from the source. In order for current to increase, the total circuit RESISTANCE MUST DECREASE. Each lamp added to the circuit in parallel will decrease the total resistance and the circuit will draw more current.

Continue your experience using four lamps in parallel. Your power company supplies a fixed voltage to your home, usually 117 volts. How are lights and appliances connected in your home? Why? If you remove one lamp in a series circuit, will the others continue to glow? Why? If you remove one lamp from a parallel circuit, will the others continue to glow? Why?

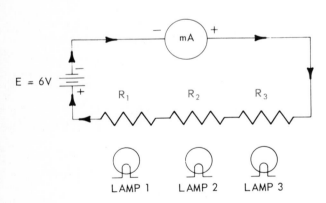

Fig. 6-15. The series circuit where L_1, L_2 and L_3 have been replaced with their equal resistances R_1, R_2 and R_3.

WHAT IS A LOAD?

In your previous experience you used lamps for a load on the power source. Other components and devices can also be used. However, regardless of the types of components connected in series or parallel, if they use POWER from the source they can be represented as a resistive load (resistance). Resistance is the only property of a circuit which uses POWER. The circuits previously used for series and parallel experiences are now drawn in Figs. 6-15 and 6-16. The lamps are replaced with resistors. The value of each resistor is equal to the resistance of the lamp it represents.

Fig. 6-16. The parallel circuit where L_1, L_2 and L_3 have been replaced with their equal resistances R_1, R_2 and R_3.

WHAT IS A SHORT CIRCUIT?

Like water flowing down a mountain, electricity will take the easiest path. Electricity also follows this law of nature. It will always flow in the PATH OF LEAST RESISTANCE. Looking at Fig. 6-17, a power

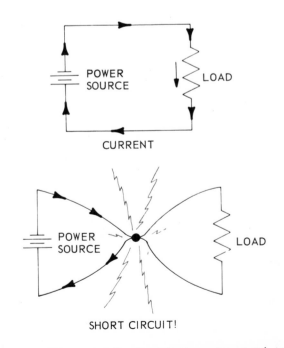

Fig. 6-17. When uninsulated conducting wires touch, the resistance is ZERO. HIGH CURRENTS – SPARKS – and HEAT result. This is a DANGEROUS CONDITION.

source is connected to a resistive load. The load limits the current. Then the uninsulated connecting wires are accidentally brought in contact with each other. ZOWIE! Sparks fly - - the wires get red hot. This is a SHORT CIRCUIT, a shorted path for the current to get around the load. This dangerous condition can cause severe burns and may set a house on fire! How many lamps and cords like the one in Fig. 6-18 do you have around your house?

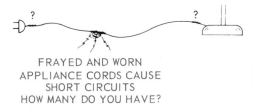

FRAYED AND WORN
APPLIANCE CORDS CAUSE
SHORT CIRCUITS
HOW MANY DO YOU HAVE?

Fig. 6-18. Frayed and worn appliance cords are a serious source of danger and fire!

Conducting wires are covered with insulating materials such as rubber and plastic. The insulation prevents unwanted and dangerous SHORT circuits between wires and components.

PROTECT YOURSELF AND YOUR PROPERTY

A FUSE is a protective device and you should be aware of just why the fuse is placed in a circuit. As you have previously studied, a SHORT CIRCUIT will cause very high currents to flow from the power source. These currents produce HEAT! The wires can get RED HOT and even MELT. Any ma-

terial such as wood or paper nearby can be set on fire! At this very moment a home belonging to a family like yours, is burning to the ground because of improper or insufficient circuit protection. How can this be prevented? A fuse is an automatic circuit protector. Common types are made of a short length of metal or wire which quickly melts and "opens" the circuit if more than a certain amount of current flows. Notice the diagram in Fig. 6-19, and the symbol for a Fuse. Notice that the fuse is connected in SERIES with the load. If the load should

Fig. 6-19. A simple circuit of power source, lamp and protective fuse. In case of a short circuit the fuse would "blow out."

draw too much current the fuse element quickly melts and the circuit no longer works. The circuit will not work again until the fuse has been replaced with a new one. BEFORE A FUSE IS REPLACED, THE REASON THAT MADE THE FUSE BURN OUT MUST BE DISCOVERED AND CORRECTED. Fig. 6-20 shows a variety of fuses used in electrical and electronic circuits.

Fig. 6-20. Various types and sizes of FUSES. A—Glass Tube Fuses used in electronic equipment and automotive electrical circuit. B—Cartridge Fuses found in home and industrial electrical circuits. C—Typical Plug Fuse found in many home electrical circuits. D—The FUSETRON which has a delay time before blowing. This allows high currents required to start a motor to flow momentarily. (Bussman Mfg. Div., McGraw Edison Co.)

A

B

C

D

When the power company brings electricity to your home, it is connected to an entrance switch and fuse box. Then the electricity is divided into several circuits to supply the lamps, convenience outlets, heaters, ranges and other devices. Each circuit will have its own individual protection. Some circuits used for lamps only will be protected with 15 ampere fuses. A circuit for kitchen appliances may have a 20 ampere fuse. Other circuits for ranges and heaters will have larger fuses. A typical ENTRANCE SWITCH and DISTRIBUTION BOX is pictured in Fig. 6-21 which uses CIRCUIT BREAKERS instead of fuses. A circuit breaker is either a thermal or magnetic switch which opens the circuit when too much current flows.

Fig. 6-21. A 100 ampere 120/240 volt Main Circuit Breaker Load center. The indicator light shows which circuit breaker is open. (Square D Co.)

SWITCHES

A switch is a simple component which will OPEN or CLOSE a circuit. A switch used with a lamp will turn the lamp ON or OFF. In the open position, there is no electrical connection between its terminals. In this case the switch would represent infinity (unlimited) resistance. In its closed position, the contacts are touching each other and there is a path of zero resistance through

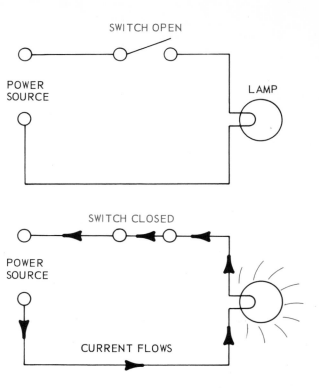

Fig. 6-22. A Single-Pole switch is used to control the current to the lamp. A SPST switch.

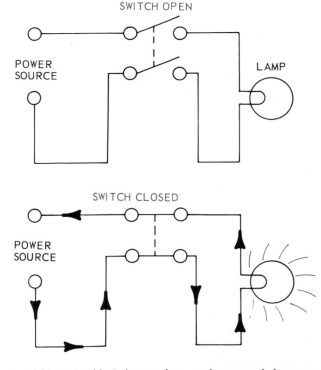

Fig. 6-23. A Double-Pole switch is used to control the current in both wires leading to the lamp. A DPST switch.

Fig. 6-24. A Single-Pole Double-Throw switch (SPDT) is used to turn either Lamp 1 or Lamp 2 on. Trace the current flow.

Fig. 6-25. A Double-Pole Double-Throw switch (DPDT) is used to switch both sides of the power line to either Lamp 1 or Lamp 2.

the switch. Fig. 6-22 shows a simple SINGLE-POLE switch used to turn the lamp ON or OFF. In Fig. 6-23 a DOUBLE-POLE switch performs exactly the same job, however the double-pole switch OPENS and CLOSES both wires leading to the lamp. This type of hookup is required under certain circumstances.

In Fig. 6-24 a SINGLE-POLE DOUBLE-THROW (SPDT) switch is used. In position

1, Lamp 1 will glow; in position 2, Lamp 2 will glow. It is an EITHER - OR situation. Both lights cannot glow at the same time. Observe the current flow indicated by arrows.

Fig. 6-25 illustrates a DOUBLE-POLE DOUBLE-THROW (DPDT) switch which could also be used if both sides of the line required switching.

ROTARY SWITCHES

Many rotary switches are used in electronic circuits. A rotary switch is a multiple

Fig. 6-26. A ROTARY SWITCH is used to turn on any desired lamp. The current for position 1 to cause Lamp 1 to glow is indicated by arrows. Other positions of switch are shown by dashed lines.

Fig. 6-27. A rotary switch with two wafers of contacts controlled by a single rotary shaft. Many of these types of switches are used in electronic equipment. (Centralab)

contact switch which may be turned to any desired position. An example is the control knob located on the front panel of a piece of equipment. In Fig. 6-26 a four position rotary switch is used. The desired lamp may be turned on by rotating the switch to the required position. Fig. 6-27 illustrates one of the commercial type switches found in electronic equipment. These are assembled in many forms to meet circuit needs.

TURN ON SEVERAL LAMPS

A load can consist of one or more lamps in either series or parallel. Since we wish all lamps to glow with full brightness, it will be only necessary to connect extra lamps in parallel with the first lamp. This may be observed in Fig. 6-28 where three lamps are turned ON or OFF at the same time by a single switch. What would be the result if the three lamps were connected in SERIES? Results: Each lamp would burn at only one

third of its required brightness. Also if one lamp should burn out, all the lights would go out. You may draw your own conclusions as to the method of connecting lamps and appliances in your home wiring circuits.

Fig. 6-28. Three lamps in parallel are controlled by one SPST switch. All lamps burn at full brightness.

THREE-WAY SWITCHING

Three-way switching is studied because of its popularity in house wiring circuits. Many times you may wish to turn a light

ON as you enter a room and turn it OFF as you leave the room from another door. You may wish to turn lights ON or OFF at either the top or bottom of a staircase. It certainly saves a lot of steps and "wear and tear" on the carpet, as well as providing personal safety by wiring light circuits so that they

Fig. 6-29. *Three-way switch connections. The lamp can be turned "on or off" by either switch 1 or 2.*

may be switched from more than one location. Fig. 6-29 shows the connections for the three-way circuit. Two switches are required. The light can be turned either ON

or OFF by either switch. A typical sketch of the three-way switch connection used in house wiring appears in Fig. 6-30. Note the three conductor cable is used from switches to junction box. Power is brought to the junction box by a two wire cable. The lamp is shown in the ON position and the complete current path is marked by colored arrows.

DECISION SWITCH CIRCUITS

In the modern computer the ON and OFF operation is not performed by mechanical switches since it is much too slow. This is accomplished by high-speed switching diodes and transistors. However, the theory of switching is the same. In computers a switch is considered to be in one of two states, either ON or OFF. This relates to BINARY (base two) numbers in which ON is represented by number 1 and OFF is represented by 0. You may also read your own answers to any specific questions using YES for 1 or switch ON and NO for 0 or switch OFF.

Fig. 6-30. *TYPICAL CONNECTIONS IN HOME WIRING FOR 3-WAY SWITCHES TO OPERATE ONE OVERHEAD LAMP. FOLLOW CURRENT PATH marked with arrows. Lamp is ON. Notice wire color code and how wires are connected. Compare with Fig. 6-29.*

With this in mind look at the decision circuit of two switches in Fig. 6-31. This circuit is known as an AND gate. This is another way of saying; IF S_1 IS YES OR ON AND S_2 IS YES OR ON THEN THE CIRCUIT IS COMPLETE AND THE LAMP GLOWS IN A POSITIVE YES. Beside the circuit is a typical TRUTH TABLE showing the four possible conditions of the circuit. The symbol for the AND gate used in computer diagrams is also shown.

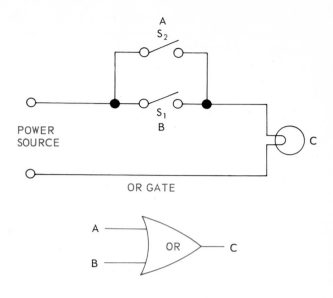

OR GATE

TRUTH TABLE

A	B	C
S_1	S_2	LAMP
1	0	1
0	1	1
1	1	1
0	0	0

0 MEANS "OFF"
I MEANS "ON"

Fig. 6-32. OR GATE SWITCHING. Notice the lamp will glow showing a complete circuit if either S_1 or S_2 is closed or in 1 position.

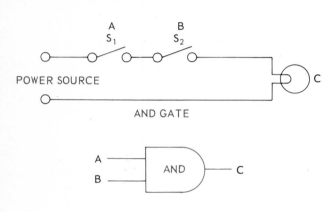

AND GATE

TRUTH TABLE

A	B	C
S_1	S_2	LAMP
0	0	0
1	0	0
0	1	0
1	1	1

0 MEANS "OFF"
I MEANS "ON"

Fig. 6-31. AND GATE SWITCHING. Notice that both S_1 AND S_2 must be in the ON or 1 position to realize an output 1 indicated by glowing lamp.

THE OR GATE CIRCUIT

The arrangement of another popular computer circuit is illustrated in similar fashion in Fig. 6-32. In this case S_1 and S_2 are in parallel and it can be stated that, IF EITHER S_1 OR S_2 IS CLOSED THERE WILL BE AN OUTPUT AND THE LAMP WILL GLOW. The TRUTH TABLE for this circuit is also shown.

Simple and very complex decisions can be made by using the circuits alone and in combination.

A COMPUTER STORY

Let's suppose you want to go to the basketball game this Friday night but your parents have set up these conditions:
A. You must do your homework.
B. You must find a friend to go with you.
C. You must promise to be home at 10 o'clock.

This is written by using logic symbols Fig. 6-33. If the three inputs A, B and C are 1 or YES, then D is YES with permission to go. If any one of the conditions is NO or 0 then D will also be NO. You will have to stay home and watch the game on TV.

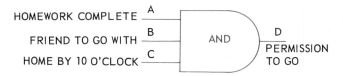

HOMEWORK COMPLETE — A

FRIEND TO GO WITH — B — AND — D PERMISSION TO GO

HOME BY 10 O'CLOCK — C

Fig. 6-33. The logic symbol for computer story in text.

INVERTERS

An inverter is added to the basic gate circuit to produce an answer opposite to the original. The symbol is a small circle illustrated in Fig. 6-34 for both the AND and OR gates. With the inverter added the gates become NAND (not and) and NOR (not or). For the NAND gate, ONES at A and B produce ZERO for output. For the NOR gate, ONES at either A or B produce ZERO for output.

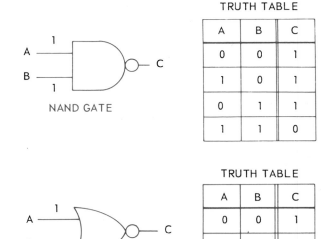

NAND GATE

TRUTH TABLE

A	B	C
0	0	1
1	0	1
0	1	1
1	1	0

NOR GATE

TRUTH TABLE

A	B	C
0	0	1
1	0	0
0	1	0
1	1	0

Fig. 6-34. Symbols and truth tables for NAND and NOR gates.

LOGIC EQUATIONS

When writing the description of gates and logic diagrams, the correct expressions must be used.

For the AND gate. An input at A and B produces output C. This is written:

$$AB = C$$

For the OR gate. An input at A or B produces output C. This is written:

$$A + B = C$$

Study the symbols and equations of the logic diagrams in Fig. 6-35. Write a TRUTH TABLE for each equation.

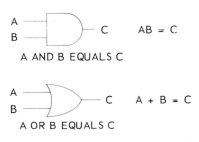

A
B — C $AB = C$

A AND B EQUALS C

A
B — C $A + B = C$

A OR B EQUALS C

A AND B OR C OR D EQUALS E $AB + C + D = E$

A AND B OR C OR D AND E EQUALS F $AB + C + DE = F$

A OR B AND C AND D OR E EQUALS F $(A + B) C (D + E) = F$

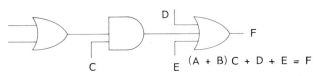

A OR B AND C OR D OR E EQUALS F $(A + B) C + D + E = F$

Fig. 6-35. Study the logic symbols and equations.

WIRE NUMBER

DIGIT NUMBER	3	2	1	BINARY NUMBERS
0	0	0	0	000
1	0	0	1	001
2	0	1	0	010
3	0	1	1	011
4	1	0	0	100
5	1	0	1	101
6	1	1	0	110
7	1	1	1	111
8	1	0	0	0	1000

Fig. 6-36. Signal or voltage level on three wires can represent binary numbers from 0 to 7. A fourth wire is needed to make an 8.

COUNTING CIRCUITS

We are used to counting by the digit system and we use numbers from 0 to 9. In the BINARY system counting is possible using only 0 and 1. Any of the familiar numbers can be represented by some combination of ONES and ZEROS. In computers the ONES and ZEROS can represent different voltage levels on a line or wire. For example, we might consider zero voltage as binary number 0 and a positive 10 volts as number 1. Consider now the chart in Fig. 6-36 in which only three wires are used. By changing the voltage from 0 to 10 or the signal from 0 to 1 on these wires, they can represent any number between 0 and 7. By adding a fourth wire we could count to 15 and so on.

Fig. 6-37. Four flip-flops connected in cascade. Each has an indicating light.

FLIP-FLOPS

A very useful circuit in the computer is the flip-flop, which usually consists of two

PULSES	DIGITAL NUMBERS	INDICATING LAMPS	BINARY NUMBERS
	0	○ ○ ○ ○	0000
	1	○ ○ ○ ●	0001
	2	○ ○ ● ○	0010
	3	○ ○ ● ●	0011
	4	○ ● ○ ○	0100
	5	○ ● ○ ●	0101
	6	○ ● ● ○	0110
	7	○ ● ● ●	0111
	8	● ○ ○ ○	1000

Fig. 6-38. The indicator lights show the binary number corresponding to input pulses.

Fig. 6-39. A commercial educational adder designed to teach computer theory and the binary number system. Pulses are produced by the telephone dial.
(EDUCOMP, Dr. Larry Heath, Oregon State University)

CONCLUSIONS

LIGHT 1   switches with EVERY PULSE
LIGHT 2   switches every SECOND PULSE
LIGHT 3   switches every FOURTH PULSE
LIGHT 4   switches every EIGHTH PULSE

This circuit is an ADDER CIRCUIT. After a problem the adder can be RESET to zero by a reset pulse. Fig. 6-39 shows an educational application of the binary adder to assist you in understanding the fundamentals of a computer and numbers systems. The number of pulses or numbers to be added are selected by using the telephone dial. For larger numbers, several adders are connected together.

transistors. When one is conducting or ON the other is OFF. A pulse of voltage will cause the circuit to flip and number one will turn OFF and the second will turn ON. Another pulse will cause the circuit to flop back to its original state. Indicating lights are used to show which transistor is conducting. The light ON represents a binary 1, the light OFF is a binary 0. In Fig. 6-37 four of these flip-flops (note the logic symbol) are connected in CASCADE. Now examine what happens when pulses are applied to the circuit input as illustrated in Fig. 6-38.

Fig. 6-40 shows a large computer center used to process business and scientific information.

FORWARD STEPS IN UNDERSTANDING ELECTRICITY-ELECTRONICS

1. An electrical circuit consists of a power source, a load, the control and connecting wires.

Fig. 6-40. This large Honeywell 8200 Computer performs business and scientific processing tasks for the Texas Eastern Transmission Company.

2. Current flow in a circuit is the transfer of energy between electrons in a conductor.

3. A poor or nonconductor is called an insulator.

4. Wires are sized by Gage Number, the larger the number, the smaller the wire.

5. Electrical energy travels at approximately the speed of light or 186,000 miles per second or 300,000,000 meters per second.

6. In a Series circuit there is only one path of current flow and the current must be the same at any point in the circuit.

7. Series connections of resistive components such as lamps, increase the total resistance of the circuit.

8. In a Series circuit the voltage divides according to the resistance of the component and the current. It is called the IR drop.

9. In a Parallel circuit there are multiple paths for current flow. Components are connected side-by-side.

10. In a Parallel circuit the total current is equal to the sum of all the branch currents.

11. Adding more resistive components in Parallel increases the total circuit current and therefore must also reduce the total resistance.

12. Only resistance as a load uses power from the source.

13. A short circuit is zero resistance and maximum current flows.

14. A fuse is a circuit protective device.

15. A switch is a control component and is used to open or close a circuit. Open means to TURN OFF. Close means TURN ON.

16. Switching circuits are the fundamental circuits of all computers.

TEST YOUR KNOWLEDGE - UNIT 6

1. What is the resistance in ohms of 100 feet of No. 20 hookup wire?

2. Measure the resistance of a foot of No. 20 Nichrome wire.

3. What would happen if you used Nichrome wire to connect up a lamp circuit?

4. In a series circuit, why do all lamps go out when one lamp is removed from its socket?

5. Why would the electrician wire your home with several independent circuits, rather than one heavy circuit?

6. All outlet plugs and ceiling lights in one room of a house should be on the same circuit. Yes or No? Why?

7. Which will use more electric power? Four lamps in series or four lamps in parallel? Explain.

8. Draw the wiring diagram for problem 7 and then prove circuit by actual connections.

POWER SOURCE

LAMP 3 REQUIRED

SPST SWITCH 2 REQUIRED

SPDT SWITCH 2 REQUIRED

PUSH BUTTON SWITCH 2 REQUIRED

DOORBELL

USE THESE COMPONENTS IN THE FOLLOWING EXPERIENCES.

EXPERIENCE A. Connect Lamp 1 only to be controlled by single-pole switch.

B. Connect Lamps 1, 2 and 3 to be turned on by single-pole switch. All lamps to burn brightly.

C. Connect Lamp 1 in a three-way switch circuit so that it may be controlled by either S_1 or S_2.

D. Connect circuit with SPDT switch to light Lamp 1 in one position and Lamp 2 in its other position.

9. Draw a wiring diagram of a doorbell circuit for your home. Use push button for front door and rear door to ring bell.

10. Draw a circuit using 3 eight-pole rotary switches to operate a lamp or special magnetic lock. The secret combination to activate the lamp or lock will be 3-5-7.

RULES OF SAFETY

FUSES ARE PROTECTION

A circuit designed to carry a certain current should have a fuse which will BLOW OUT if the circuit current is exceeded. Always investigate the cause that burned out the fuse. Do not replace a fuse until the cause of its BLOW OUT is found. Never replace with a larger fuse or a jumper. You are inviting trouble.

OCTOPUS EXTENSIONS

Older homes frequently have too few wall receptacles to serve all the lamps and appliances required for modern daily living. The use of multiple 3-way or more extension cords is dangerous. Wires may get hot and burn! You may get some severe shocks!

FIRST AID

All well-run shops and laboratories have FIRST AID kits in handy locations. Know where the first aid kit is in your shop. Report any accident to your teacher, no matter how minor. He will supply treatment where needed.

DON'T BREAK YOUR BACK

Some electronic equipment is heavy, especially the types which use transformers. Always get help to lift or move it around. Ask your teacher to demonstrate the proper way to lift a heavy object.

HOT SOLDERING IRONS

A hot soldering iron or gun appears innocent. If left around on the bench or hung in the tool cabinet, some unsuspecting fellow student may put his hand on it. Don't let this happen.

LONG CORDS DROP VOLTAGE

A long extension cord has resistance and the voltage at its end may be somewhat less than required for an appliance or motor. It may shorten the life of your tools and appliances to operate them with reduced voltage.

SAVE YOUR SIGHT

Your eyesight is a valuable gift of life. Protect your eyes when grinding, chipping or polishing. Also see that your buddy nearby is protected.

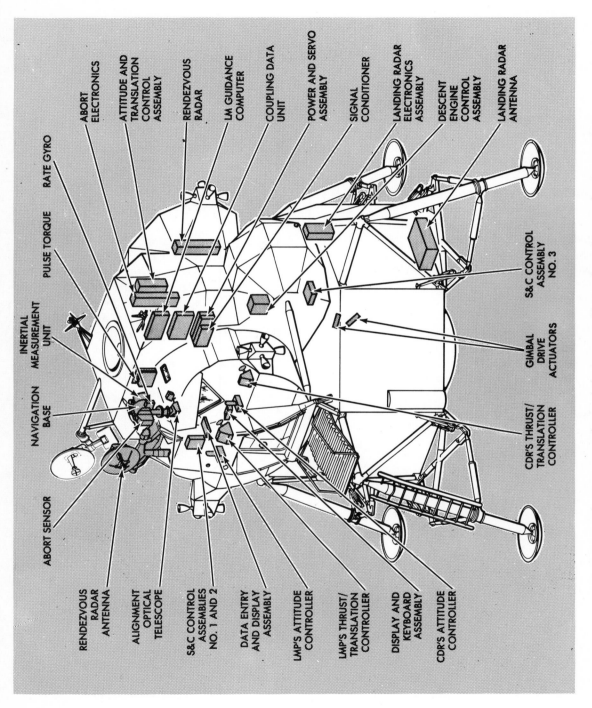

RENDEZVOUS RADAR ANTENNA

ALIGNMENT OPTICAL TELESCOPE

S&C CONTROL ASSEMBLIES NO. 1 AND 2

DATA ENTRY AND DISPLAY ASSEMBLY

LMP'S ATTITUDE CONTROLLER

LMP'S THRUST/ TRANSLATION CONTROLLER

DISPLAY AND KEYBOARD ASSEMBLY

CDR'S ATTITUDE CONTROLLER

ABORT SENSOR

NAVIGATION BASE

INERTIAL MEASUREMENT UNIT

PULSE TORQUE

RATE GYRO

ABORT ELECTRONICS

ATTITUDE AND TRANSLATION CONTROL ASSEMBLY

RENDEZVOUS RADAR

LM GUIDANCE COMPUTER

COUPLING DATA UNIT

POWER AND SERVO ASSEMBLY

SIGNAL CONDITIONER

LANDING RADAR ELECTRONICS ASSEMBLY

DESCENT ENGINE CONTROL ASSEMBLY

LANDING RADAR ANTENNA

S&C CONTROL ASSEMBLY NO. 3

GIMBAL DRIVE ACTUATORS

CDR'S THRUST/ TRANSLATION CONTROLLER

Astronauts guide, navigate, and control the LM using 55 switches, 45 circuit breakers, and 13 indicators which are tied to the rest of the ship. This equipment is contained in a primary guidance and navigation section, an abort guidance section, a control electronics section, and in the landing and rendezvous radars. (NASA)

Unit 7

THE MYSTERY OF MAGNETISM

Legends handed down by word of mouth for centuries told the exciting story of a mysterious kind of stone found on the earth's surface in Asia Minor. The "living stone" which would share its life with other stones and attract small pieces of iron became a matter of deep concern for the scientist and the philosopher in the dark ages.

It was not until the 12th century that English scientists decided this suspicious stone did not contain life, but in some manner was related to another mysterious force called "electricity."

In the early 17th century Sir William Gilbert proposed the almost unbelievable theory that the earth on which we live is an enormous magnet. He found that a small living (magnetic) stone suspended on a string or placed on a small wooden float in a vessel of water would turn in a NORTH-SOUTH direction. Early navigators realized the value of the living stone and used it as a direction finder when sailing their ships in unknown seas.

The natural magnet has been replaced in the electronic age by more dependable and stronger permanent magnets made of steel and alloys. Much has been learned about the magnetic field of force existing in space around the surface of the earth.

VISIBLE MAGNETIC FIELDS

Although you cannot look into space and see a magnetic field, such a field does exist in space all about us. The effect of this field can be seen by performing the learning experience shown in Figs. 7-1 and 7-2. Place a sheet of cardboard over a single magnet. Fine iron filings are then sprinkled on the cardboard, much like you would sprinkle

CARDBOARD

Fig. 7-1. *IRON filings arrange themselves to form a pattern of the magnetic field.*

Fig. 7-2. *The magnetic field can be seen and photographed by using this demonstration device.*
(Lab–Volt, Buck Engineering Co.)

Fig. 7-3. The lines of force combined when the magnets are placed in the attractive position.

salt on a hamburger. Now gently tap the cardboard and the iron filings will arrange themselves along the "lines of magnetic force" between the North and South poles. Some observations should be made while studying these lines of force.

1. Each line is an individual line and no two lines cross each other.
2. The lines are dense at each pole and enter the pole at a perpendicular angle.
3. Each line forms a continuous path from North to South through the magnet to North. There is no movement within this path.

LAWS OF MAGNETISM

Place two small magnets close together. If UNLIKE POLES are next to each other, the magnets will jump together. UNLIKE POLES ATTRACT EACH OTHER.

Turn the magnets so that LIKE POLES are next to each other. The magnets will jump apart. In fact, force is required to bring them even close to each other and they will jump apart if the force is removed. LIKE POLES REPEL EACH OTHER.

Fig. 7-5. The magnetic fields of opposing magnets with LIKE POLES next to each other.

Fig. 7-4. The photograph of the magnetic field when two magnets are in an attractive position.

What happens to the lines of force when two magnets are placed in the attractive position with unlike poles near each other? This is illustrated in Figs. 7-3 and 7-4. Notice the dense magnetic lines in the center between the two magnets.

Magnets placed in the opposing position with LIKE poles together and their magnetic fields are shown in Figs. 7-5 and 7-6. Notice that the space between the magnets is vacant. Magnetic lines oppose each other and will not come close.

A MAGNET AND A NAIL

Why does a magnet pick up a nail? When the nail is close to the magnet some of the

Fig. 7-6. A photograph of opposing magnetic fields.

magnetic lines will pass through the nail and cause the nail to become a MAGNET with poles necessary to cause attraction. Study Fig. 7-7. The nail, the magnetic fields and the poles produced will explain this action.

Fig. 7-7. The nail is attracted because the field of the magnet causes the nail to become magnetized.

Interesting experiments and fascinating movements of magnets are produced by the device shown in Fig. 7-8. It is called "KRAZY LEG." The magnets may be arranged to suit your own fancy.

THE CRAZY COMPASS

In the early part of the 19th century the Danish scientist, Hans Christian Oersted, was in his laboratory performing some simple experiments with electricity. A compass was nearby on the bench. Each time he turned on the electric circuit, the compass would swing on its pivot. The compass would also act up when the circuit was turned off.

Fig. 7-8. This wild game can keep you interested for hours. The magnets on the base can be arranged either attracting or opposing the magnet on the swinging leg.
(Krazy Leg, Rathcon, Inc.)

Thus, Mr. Oersted correctly assumed that his electric circuit must be affecting the compass. The long sought relationship between electricity and magnetism was discovered.

MAGNETIC FIELDS AND CONDUCTORS

Any wire or conductor carrying an electric current will create a magnetic field around the conductor and at right angles to the flow of current. This field of force forms concentric (having a common center) rings around the wire, like doughnuts strung on the wire. Fig. 7-9 illustrates these fields by dashed lines and arrows. The direction of

Fig. 7-9. A current carrying conductor has a magnetic field at right angles to current flow.

the magnetic field depends on the direction of the current flow.

LEFT-HAND RULE

The direction of the magnetic field caused by a current may be easily discovered by using the LEFT-HAND RULE. Grasp a wire with your left hand with your thumb extended outward pointing in the direction of current flow. Your fingers around the wire point in the direction of the field. The end view of a conductor and its magnetic fields may be observed in Fig. 7-10.

CURRENT
FLOWING OUT

CURRENT
FLOWING IN

Fig. 7-10. The dot in the center of conductor is the point of an arrow meaning current is flowing toward you. The cross on the end of the conductor represents the feathers on an arrow as it means the current is flowing away from you. Note the directions of the magnetic fields.

ONE FROM MANY FIELDS

Fig. 7-11 shows a wire wound into a coil. When current passes through the coil magnetic fields are formed around the wire as indicated by small arrows. These join and reinforce each other and the coil has a polarity like a MAGNET. This coil, since it has a hollow core, is called a SOLENOID. Which end is North or South? Again you may use your LEFT-HAND RULE. Grasp the coil with your left hand with your fingers encircling the coil and pointing in the direction

Fig. 7-11. The magnetic fields of each turn of the coil join together and form a coil with magnetic poles.

of current flow. Your extended thumb will point to the NORTH end of the coil.

EXPERIENCE 1. Prove that a solenoid does have magnetic polarity and the polarity changes with the direction of current.

Connect a coil to a 6 volt supply as illustrated in Fig. 7-12. Place a compass close to one end of the coil. Observe the compass action. Reverse the connections to the power source. Does the compass change direction also? Explain this by using the LEFT-HAND RULE. Does the compass point toward the North or South pole?

SOLENOID SUCKING COIL

A very interesting and useful effect of magnetism is demonstrated by the SOLENOID SUCKING COIL. Such an electrical device is widely used in home appliances and industrial machinery, whenever it is necessary to operate some mechanical switch or valve by electric power.

EXPERIENCE 2. Set up a coil and attach it to the power source as shown in Fig. 7-13. Place an iron core into the coil part way. Turn on the power. What happens? The core

POWER
SOURCE

COIL

COMPASS

Fig. 7-12. When the coil is energized, the compass indicates the polarity of the magnetic field.

Fig. 7-13. Core moves inward by SUCKING ACTION when coil is energized.

is SUCKED into the coil and comes to rest at the center of the coil.

CONCLUSION: When the coil is energized it produces a magnetic field which also magnetizes the iron core in the same manner as a magnet picks up the nail in Fig. 7-7. The core is attracted to the coil and they snap together. Once the core is in the center of the coil the magnetic field is concentrated within the core and there is no room for further movement.

EXPERIENCE 3. Set up the same solenoid coil and attach an elastic band or spring to the end of core, as shown in Fig. 7-14. Each time the circuit is closed by pushing button S_1, the core is sucked into the center. When power is removed, the core returns to its original position.

OBSERVATIONS: It seems that, in the solenoid, we have discovered a convenient method of converting electric power into

straight line motion. The moving core could be used for many purposes. It could turn a water faucet on or off. It could be used to strike a bell or chime such as found in a door chime. Refer to Fig. 7-15. It could be attached to a lever for shifting gears of a

Fig. 7-15. The solenoid is used in the simple door chime. When the coil is energized, the core is sucked in and strikes the bell. The spring causes the core to return to its original position.

Fig. 7-14. The elastic band returns core to its original position when power is removed.

machine. It might be used to activate a bolt lock on a door. Can you think of other ways to use a SOLENOID?

ELECTROMAGNETS

An improvement can be made on the strength of the magnetic field produced by the solenoid. An iron core is placed in the center of the coil and it is now called an ELECTROMAGNET.

Fig. 7-16. *The electromagnet picks up iron nails like the permanent magnet. An increase in voltage causes an increase in current and an increase in the strength of the magnet.*

EXPERIENCE 4. Assemble the coil and the core as instructed in Fig. 7-16. Connect it to your power source and adjust the voltage to about five volts. Check the coil with the compass to make sure a magnetic field does exist. Pick up several small nails or paper clips with the electromagnet. It works like a permanent magnet, does it not?

Increase the voltage of your power source to ten volts. Notice that the magnet becomes much stronger and will pick up more nails.

CONCLUSIONS: An increase in voltage

causes an increase in current and therefore an increase in magnetic field strength. The strength of an ELECTROMAGNET is measured in AMPERE TURNS abbreviated to IN where I is the current in amperes through the coil and N is the number of turns of wire in the coil. A coil with 200 turns and one ampere of current would have a strength of 200 IN.

$$1 \times 200 = 200 \text{ IN}$$

If current is increased to 2 amperes then the field strength would be 400 IN.

$$2 \times 200 = 400 \text{ IN}$$

MAGNETIC MATERIALS

There are only a few materials which will be attracted by a magnet, such as iron, nickel and cobalt. We have been using iron for a core in our experiences.

EXPEREINCE 5. Secure some samples of other kinds of materials like wood, plastic, zinc and copper. Try to pick up these materials with the electromagnet. The magnet does not work because these are not magnetic materials.

A boat builder went to a hardware store to purchase some brass screws which do not rust in water. The hardware dealer tried to sell him some brass plated screws. The boat builder had brought a small magnet with him. How did he discover that the screws were not brass? Hint: Brass is an alloy of copper and zinc.

TEMPORARY AND PERMANENT MAGNETS

Set up your experience for an electromagnet again as in Fig. 7-16. After the electromagnet has been energized, turn off the power and remove the core. Test the magnetism of the core by trying to pick up some nails. You will find the magnet is very weak.

CONCLUSION: The core is made of soft iron with a very low carbon content and does not retain magnetism after the force which magnetized it has been removed.

EXPERIENCE 6. Repeat the activity, but this time use a steel core. Notice that the steel will pick up nails after the magnetizing force has been removed.

CONCLUSION: Some materials, such as steel, will retain the magnetism after they have been magnetized.

Frequently when working on electrical circuits with a steel screwdriver, the screwdriver will become magnetized. This can be a nuisance, since it attracts particles of iron and steel and loose nuts and bolts.

MIRACULOUS FERRITES

You should know something about some new materials made of iron oxide compounds. These are called FERRITES. These magnetic materials have almost, single-handedly, revolutionized the electronic industry. Without them, the development of television, computers, and audio and video recorders would not have been possible.

A ferrite is an electrical insulator and can be magnetized in either a North-South or South-North polarity by applying a very small magnetizing force. Consequently, "ferrite memory cores" are widely used in the memory storage of the modern computer. The polarization can be related to the BINARY NUMBER SYSTEM. If the core is magnetized in one direction, it represents a ONE; if magnetized in the opposite direction, it can mean a ZERO. Therefore hundreds, actually millions, of numbers and symbols can be stored by using the ferrite cores in a memory plane. A simple example is illustrated in Fig. 7-17 which is a plane of horizontal and vertical wires. At each intersection is a tiny doughnut of ferrite material. A current in any one horizontal and vertical wire will magnetize one and only one core. Any core can be magnetized by selecting the right combination of wires. See photo in Fig. 7-18.

Since the core remains magnetized, it will store that information. The computer can retrieve the information by sensing the

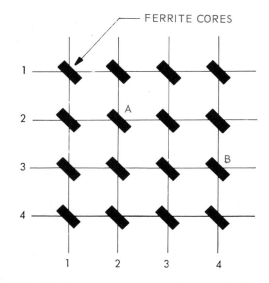

Fig. 7-17. Ferrite cores are strung on horizontal and vertical wires. A core can be magnetized by current through any selected pair of wires. Note 2-2 magnetizes core A and 4-3 magnetizes core B.

magnetism of the core. You will want to study this later in courses in electronics.

Ferrites are also used on audio and video recording tapes. The music or picture is converted into a varying electric current which magnetizes the tape in that particular pattern. For play-back, a sensing head interprets the pattern and produces an electrical potential corresponding to the recorded pattern - - Presto! A tape recorder.

THE MAGNETIC SWITCH

The RELAY or magnetic switch finds thousands of uses in electronic communications and industrial controls. It provides fast and positive switching for electrical circuits. It assures safety for workmen because machines can be controlled from a remote location. Relays are economical to use, since only large and expensive wires are used near the machine and the switching can be done by using low voltage and small wires.

Fig. 7-18. Ferrite cores (doughnut-shaped pieces of metal) strung on a wire grid provide the mechanism for recording "bits" (binary digits) of data in the main memory of a computer. (Honeywell, Inc.)

A RELAY consists of four main parts. These can be seen in Fig. 7-19. They are the switching contacts, the armature, the electromagnet and a spring.

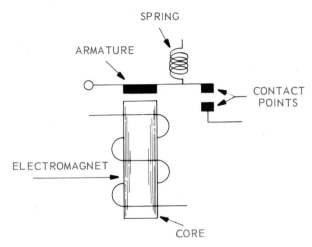

Fig. 7-19. *The main parts of a* **RELAY.**

When a voltage is applied to the electromagnet, the magnetism attracts the armature and pulls it downward causing the contact points to close. When the coil is demagnetized, the spring pulls the contact points apart and they return to their normal position.

EXPERIENCE 7. The circuit for a relay is shown in Fig. 7-20. Note the schematic symbol used to represent a relay. Also note that this relay has another contact point. The relay can be used to either turn a circuit ON or OFF depending on which contact points you wish to use. This circuit is arranged to turn the lamp ON.

Now rearrange the lamp connection so that the lamp is normally ON and will go OFF when you push the button.

How much voltage is required to operate the relay? Of course you realize the lamp could be a motor or some other high voltage or high current device.

EXPERIENCE 8. This time we will arrange the circuit using two lamps. L_1 will be ON and L_2 will be OFF. Study Fig. 7-21. When the button is pushed, L_1 will go out and L_2 will turn ON. Is there any ELECTRICAL connection between the input or controlling circuit and the OUTPUT or controlled circuit? This is one of the major advantages of relay circuits.

CONCLUSION: The input circuit could be only a few volts and use small inexpensive

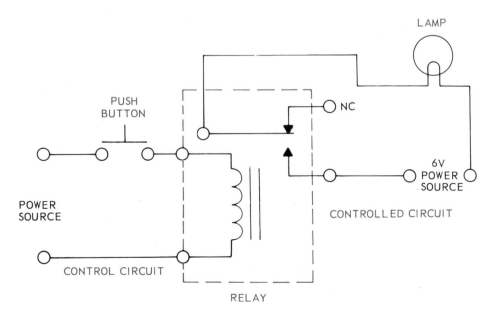

Fig. 7-20. *The* **RELAY CIRCUIT.** *When the button is pushed, the coil is energized and the armature is attracted downward. The points are* **CLOSED** *and the lamp glows. NC means "no connection."*

Fig. 7-21. *Relay circuit* L₁ *is* **ON**, L₂ *is* **OFF**. *When button is pushed* L₁ *goes* **OFF**, L₂ *is turned* **ON**.

wires, while the output or controlled device could require several hundred dangerous volts, large wires and heavy currents. Of course, the push button switch could be miles away and the operator would be safe from any dangers.

A BUZZER

A buzzer or doorbell uses electrical principles similar to the relay. The difference is in how the connections are made.

Fig. 7-22. *The circuit for a* **BUZZER.**

EXPERIENCE 9. A circuit for a BUZZER is drawn schematically in Fig. 7-22. An interesting experience for you would be to build this buzzer and observe its action.

CONCLUSIONS: When the coil is energized by pushing the button, the armature is attracted downward causing the points to open. This opens the circuit. The coil magnetism drops to zero and the contact points are allowed to close. This again energizes the circuit and the points open. This action continues as long as a voltage is applied to the buzzer, so called, because it buzzes.

CHOPPER

Consider for a moment the current flowing through the buzzer circuit. It is not a continuous current. In fact, when the contact points open, the current drops to zero. When the points are closed, the current is maximum. So the current is rising and falling as

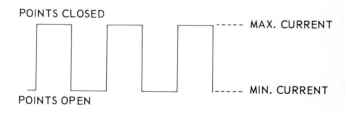

Fig. 7-23. *This square wave shows the action of the buzzer-chopper in a circuit.*

the armature vibrates. Fig. 7-23 shows a simplified graph of the current in the circuit. It looks like a SQUARE WAVE.

Up to this point in your studies, only DC or DIRECT CURRENT has been discussed. Direct current remains at a CONSTANT VALUE depending on the resistance of a circuit. The CHOPPER produces a PULSATING direct current. You will recall that a steady direct current produces a magnetic field. A pulsating direct current produces a PULSATING MAGNETIC FIELD. This principle is extensively used in power supplies, automotive ignition systems and electronic measuring circuits.

FORWARD STEPS IN UNDERSTANDING ELECTRICITY-ELECTRONICS

1. A magnet has a magnetic field and a NORTH and a SOUTH pole.
2. Like magnetic poles repel each other.
3. Unlike magnetic poles attract each other.
4. A magnetic field exists around a current carrying conductor. The field is at right angles to the flow of current.
5. The direction of the magnetic field around a current carrying conductor may be found by using the Left-Hand Rule.
6. A coil without a core is called a solenoid.
7. A coil with a core is called an electromagnet.
8. The strength of an electromagnet is measured in ampere-turns (IN).
9. A permanent magnet is made by placing a material which retains its magnetism in a strong electromagnetic field.
10. A relay is a magnetic switch.
11. Ferrites are iron compounds. They can be easily magnetized and demagnetized.
12. Ferrites are used for core memories in computers.
13. Direct current may be steady or pulsating.
14. A buzzer operates on magnetic action and "make and break" contact points.

TEST YOUR KNOWLEDGE - UNIT 7

1. Secure a 6 volt lamp, a relay, a switch, about ten feet of double conductor wire.

Use an available power source. Build a remote control circuit to turn a lamp on by a relay, using a switch at a distance of ten feet.
2. Secure a permanent magnet and some nails. Prove that a magnet will attract nails. Place a piece of window glass between the magnet and the nails. Does the glass affect the magnetism? Place a piece of copper between the magnet and the nails. Does copper allow the magnetic field to pass?
3. Connect an electromagnet to a power source with an ammeter in series. Adjust voltage to five volts. What is the current flow? Secure another iron core. Increase voltage until the electromagnet will pick up and hold the second core. Decrease the voltage slowly watching the current meter. At what current is the magnetic field sufficiently weak to drop the second core? What does this experience prove to you.
4. Wire up a doorbell or buzzer circuit with push buttons at the front and rear door. Use your relay as a buzzer. Use two push buttons and two pieces of double conductor wire. Either button must ring the buzzer.
5. Build an overcurrent indicator. Connect circuit by schematic Fig. 7-24. Notice that the relay is in series between power

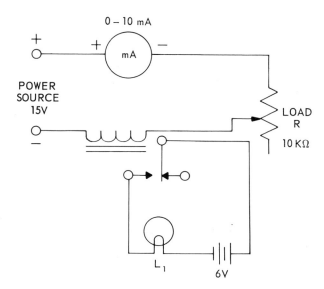

Fig. 7-24. Circuit for problem 5.

source and load. Set voltage at 15 volts. Adjust load resistance for "closing current" of relay. At this point L_1 will glow. Current = _____ .

Increase load R and reduce current until relay opens. Minimum holding current = _____ .

6. Construct the relay buzzer circuit of Fig. 7-22. Ask your instructor to connect an oscilloscope across coil. Draw a picture of the wave form.
7. Make a list of five ways that a solenoid might be used as an electrical-mechanical control.

Fig. 8-1. A laboratory setup for performing the experience of generating electricity.

Unit 8

DIRECT AND
ALTERNATING CURRENT

In our study of Sources of Electrical Energy the use of a magnetic field and a moving conductor was purposely delayed. This important source accounts for most of the electrical power now used in homes and industry. The development and growth of our industrial and military giants can be attributed to a discovery by Michael Faraday early in the 19th century. This discovery showed a definite relationship between a magnetic field and electricity. Faraday is remembered in history as the "father of the dynamo."

This Unit can be one of your more important studies of electricity. To assure your understanding, let's perform and "rediscover" the principles which made Faraday famous:

EXPERIENCE 1. A coil is connected directly to a milliammeter as in Figs. 8-1 and 8-2. The milliammeter used must have a pointer which can move either to the right or left of zero. Take a magnet and hold it above the hollow core of the coil. Watch the meter as you quickly push the magnet into the coil. Which way does the meter move? To the right or to the left? Now quickly with-

MILLIAMMETER

MAGNET

COIL

Fig. 8-2. Coil is attached to milliammeter to demonstrate the generation of electricity.

draw the magnet from the core. Which way does the meter move? To the right or to the left? Does the meter indicate any current in either direction when the magnet is not moving?

CONCLUSIONS: We have a coil of conductors and a magnetic field close to each other. Nothing happens unless there is movement of the magnet. Movement of the magnet downward into the coil produces a current, as read on the meter, in ONE DIRECTION. Pulling the magnet out of the coil produces a current in the OPPOSITE DIRECTION. You have made a simple GENERATOR.

EXPERIENCE 2. Repeat Experience 1 with these changes. Hold the magnet in a fixed position and MOVE THE COIL DOWN OVER THE MAGNET. Does the meter pointer move to the right or the left? Pull the coil from the magnet. Does the meter pointer move to the right or left? Does the meter move at all when there is no movement of the coil?

CONCLUSIONS: Electricity can be generated by movement of either the magnetic field or by moving the coil. The DIRECTION OF THE CURRENT DEPENDS UPON THE DIRECTION OF MOVEMENT.

FARADAY'S DISCOVERY

A generator produces electrical energy from mechanical energy. There must be a conductor, such as the coil. There must be a magnetic field. There must be relative motion between the conductor and the field. This motion is mechanical motion.

ALTERNATING CURRENT (ac)

A more technical description of the foregoing experience would be: A voltage is INDUCED in the coil as it moves through a magnetic field. INDUCE means "to bring on or about." Consider these conditions from your experience:

No movement results in zero voltage and zero current.

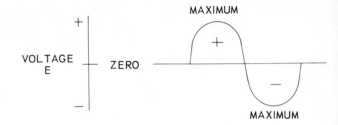

Fig. 8-3. *This illustration shows the induced voltage and polarity during one cycle.*

Downward movement results in voltage of one polarity and current in one direction.

Upward movement results in voltage of opposite polarity and current in opposite direction.

The induced voltage is illustrated in Fig. 8-3. The curve represents the magnitude or greatness of the voltage and the polarity of the voltage during one cycle of events. A CYCLE is an interval of time in which one round of events is completed. This cycle of events is: Starting with magnet out of coil, pushing it into the coil, withdrawing magnet from coil and back to the starting point.

CONCLUSION: During the cycle the polarity of the voltage and the direction of current flow will change. If you continued to move the magnet in and out, an ALTERNATING VOLTAGE and CURRENT would be produced. Fig. 8-4 shows the resulting voltage induced during four cycles.

FREQUENCY

Frequency simply means how often the cycle of events happen. Assuming continuous

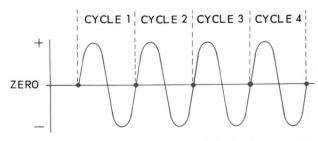

Fig. 8-4. *The induced voltage and polarity for four complete cycles.*

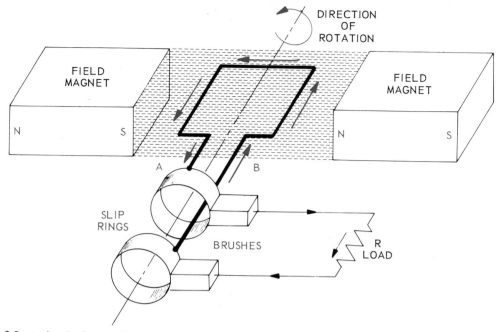

Fig. 8-5. A sketch of a simple alternator or a-c generator. Note current directions as indicated by arrows.

movement between the magnetic field and the coil, if the one cycle was completed in every second, the frequency would be:

ONE CYCLE PER SECOND or ONE HERTZ.

HERTZ is the unit of measurement for frequency and means "cycles per second." The electricity we use at home alternates 60 cycles per second or 60 Hz (note abbreviation). Sound waves such as we hear from a Hi-Fi STEREO will have frequencies from 20 to 20,000 Hz or 20 KHz (note use of prefix kilo).

Radio waves used in the AM broadcasting band have frequencies between 550 KHz and 1600 KHz. Look at the dial on a transistor radio and you will see these numbers. On a FM radio, you'll notice the dial reads from 88 MHz to 108 MHz (M means mega or one million).

A TV tuned to Channel 4 uses the frequencies between 66 and 72 MHz. Channel 12 uses 204 to 210 MHz.

SIMPLE GENERATOR

From the practical point of view, pushing a magnet in and out of a coil would be a bit

tiresome. It is better to use a fixed magnetic field and rotate a coil in the field as described in Fig. 8-5. Only a single turn of wire is illustrated for ease in understanding. The ends of the coil are brought out to copper SLIP RINGS. BRUSHES rubbing on these slip rings bring the generated electricity to an external circuit and load. Coil side A moves downward and side B moves upward as the coil rotates through its first one-half revolution. Current flows as indicated by arrows. During the second half revolution side B moves downward and side A moves upward through the magnetic field. As a result the induced voltage and current reverses direction. The output per revolution or per cycle is similar to Fig. 8-3. This machine is called an ALTERNATOR and produces an alternating current.

ALTERNATOR

From the electrical point of view it would be better not to run the generated currents through sliding contacts like the brush and slip ring. Therefore, the practical alternator generally has the coils in a fixed position and rotates the magnetic field. The fixed coils are called the STATOR and the windings of the revolving magnetic field are the ROTOR windings. The ROTOR does have slip rings, but they are connected to a direct current power source and only serve to excite the magnetic field. An ALTERNATOR from a modern automobile is shown in Fig. 8-6.

section B is with brush A and section A is with brush B. The current, therefore, continues to flow in one direction even though it has alternated in the armature. The output is a pulsating direct current.

Fig. 8-6. Cutaway view of the ALTERNATOR used on a modern car. Study the various parts described in the text. The purpose of the diodes will be discussed later under rectification. (Delco—Remy)

D-C GENERATOR

All generators produce an alternating current within their rotating armature. In order to realize a direct current output the ac output must be RECTIFIED. This means it must be changed to dc. The d-c generator uses a mechanical switch called a COMMUTATOR for this purpose. Study the diagram in Fig. 8-7.

Follow the action. Brush A is in contact with commutator section A. Also brush B and section B are together. Current through the first half revolution flows as indicated by arrows. When the current in the armature reverses during the second half revolution, the commutator has also revolved and now

GENERATOR OUTPUT

To improve upon the simple generator and increase its output, several things are done.

1. A stronger magnetic field is produced by using electromagnets. This increases the output.
2. Several coils of many turns of wire are used in the armature together with the required commutator sections. This increases the output and makes it less pulsating.

GENERATOR OUTPUT REGULATION

The speed of rotation of the generator or alternator does affect the output. However,

Fig. 8-7. A sketch of a d-c generator. The commutator switches the generator output so that current in the output circuit flows only in one direction.

it is not always convenient to change the speed. The output of these machines, both current and voltage, is regulated by varying the strength of the magnetic field. Mechanical relays and transistors are used to electronically switch more or less resistance in the field circuit and vary the field current. This varies the field strength and the generator output. Automobiles have such current-voltage regulators. An example is shown in Fig. 8-8.

Fig. 8-8. A typical relay type voltage regulator to control generator output. (Delco—Remy)

Fig. 8-9. Photographed is a transformer bank of four *General Electric* single-phase auto transformers. *(Pacific Gas and Electric Co.)*

Fig. 8-10. These circuits illustrate the changing magnetic field of a coil as an a-c voltage is applied to it.

TRANSFORMERS

The transformer is an electrical component which transfers electrical energy from one circuit to another by using magnetic fields. You will find several of these devices in most any kind of electronic equipment. Power companies use transformers to increase the voltage for their high tension cross-country power lines. The current is then reduced by transformers before the power enters your home. Fig. 8-9 shows the large transformers used by a power company at a power distribution center. You no doubt have seen such installations around the countryside. These are enclosed by high wire fences for your protection. Dogs, wild animals and people should treat these forbidden areas with respect.

Now to discover how a transformer works. Remember Faraday's discovery: To induce a voltage there must be a coil, a magnetic field and RELATIVE MOTION between the coil and the field. Study Fig. 8-10 in which a coil is attached to an ALTERNATING CURRENT source. We already know that an increased voltage will increase the magnetic

Fig. 8-11. Sketch and symbol for a transformer showing primary and secondary windings.

field around the coil. At point A of the input voltage wave, the voltage is zero and there is no magnetic field. As voltage increases to point B, the magnetic field also moves outward or EXPANDS to its maximum strength at a definite polarity. As voltage decreases to point C and zero, the magnetic field moves inward or collapses to zero. At point D the voltage again builds up to maximum in the OPPOSITE DIRECTION and the magnetic field again EXPANDS outward to maximum strength but with OPPOSITE POLARITY. At point E the voltage has returned to zero and the field has collapsed to zero. A CONTINUING A-C VOLTAGE WOULD PRODUCE A CONTINUING EXPANSION AND COLLAPSING MAGNETIC FIELD. Therefore, we have the MAGNETIC FIELD required in Faraday's discovery AND we have MOVEMENT OF THE MAGNETIC FIELD (outward and inward) produced by using an

a-c voltage. All that remains are the necessary conductors.

To meet this need a second coil of wire is placed close to the first coil. It is usually wound directly over the first coil. Both coils are wound on a common laminated iron core. Laminated? This means the core, rather than being solid, is made up of thin layers of core material. Plywood is a good example of laminating. Lamination reduces losses in the transformer.

The symbol for the transformer with its two coils and laminated core is given in Fig. 8-11. Typical types of power transformers used in electronics are illustrated in Fig. 8-12. The input coil is the PRIMARY WINDING and the output coil is the SECONDARY WINDING.

AL Case   A Case   BC Case   BS Case

Fig. 8-12. Typical types of power transformers used in electronics. (Triad)

87

ACTION IN THE TRANSFORMER

When the primary coil is connected to an a-c source of power, the expanding and collapsing magnetic field CUTS ACROSS THE SECONDARY WINDING and INDUCES A VOLTAGE IN THE SECONDARY. The output voltage depends on the number of turns of wire in the primary in respect to the number of turns in the secondary. This is called the TURNS RATIO.

Neglecting losses, if both primary and secondary have the same number of windings the ratio is 1 to 1 and the output voltage is the same as the input voltage.

STEP-UP TRANSFORMER

By increasing the turns of the secondary, a higher output voltage may be produced. For example: Assume the primary has 500 turns and the secondary 2000 turns. The ratio is now 1 to 4. An input voltage of 100 Vac would produce an output of 400 Vac. This is a STEP-UP transformer and is used to change a voltage to a higher voltage to meet the needs of certain equipment.

STEP-DOWN TRANSFORMER

As you might expect, if there are less windings on the secondary than on the primary the output voltage is lower than the input voltage. For example: Assume again that the primary has 500 turns, but the secondary has only 100 turns. Now the ratio is 5 to 1. An input voltage of 100 Vac would produce 20 Vac in the secondary or output. These STEP-DOWN transformers also are extensively used in electronic circuits.

TURNS RATIO

A simple formula may be set up by which you can always find the turns ratio or the output voltage.

$$\frac{E_{in}}{E_{out}} = \frac{N_{primary}}{N_{secondary}}$$

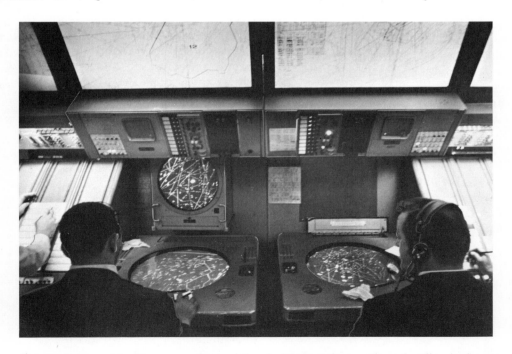

Circuits for air safety. Every second around the clock two airplanes either take off or land somewhere in the United States. This photograph shows the work station at an air traffic control center, where planes are tracked from the time of take off until they reach a point within 15 or 20 miles of their destination, where airport control takes over. All this time the various pilots get instructions from controllers who keep them in flight paths that are safely away from other aircraft.
(Bell System)

In this formula:

E_{in} = input voltage

E_{out} = output voltage

$N_{primary}$ = number of turns in primary coil

$N_{secondary}$ = number of turns in secondary coil

Work out the previous examples using this formula.

For the step-up transformer:

$$\frac{100}{E_{out}} = \frac{500}{2000} \qquad E_{out} = 400\,Vac$$

For the step-down transformer:

$$\frac{100}{E_{out}} = \frac{500}{100} \qquad E_{out} = 20\,Vac$$

The symbol for a power transformer which has both step-up and step-down windings is shown in Fig. 8-13. These transformers are designed to meet the requirements of a certain electronic circuit.

Fig. 8-13. The symbol of a transformer with both step-up and step-down windings.

Transformers are also used to transfer power from one stage of a circuit to another. This is called transformer COUPLING.

POWER TRANSFORMERS

Back in Unit 4 the very important subject of power was discussed. You should remember that:

$$POWER\ (in\ watts) = VOLTAGE\ \times\ CURRENT$$

With this in mind the circuit for a step-up transformer will be found in Fig. 8-14. In this case a load of 500 ohms is attached across the secondary. A current will flow. Assuming 100 Vac input and a 1 to 5 step-up transformer, the secondary voltage is 500 volts. The current in the secondary equals:

$$I_{secondary} = \frac{E_{secondary}}{R\ ohms}$$

or

$$I = \frac{500\,V}{500\,\Omega} = 1\ amp.$$

Fig. 8-14. The power output must equal the power input of a transformer. Losses are not considered.

The power used by the secondary therefore is:

$$I_{secondary} \times E_{secondary} = P_{secondary}$$

or

$$1\,A \times 500\,V = 500 \text{ watts}$$

Of course, if the secondary circuit is using 500 watts of power, then the PRIMARY must supply 500 watts of power. A TRANSFORMER IS NOT A SOURCE OF POWER. Then 500 watts of power is drawn from the source. But the voltage of the primary is only 100 volts. How much current must flow in the primary to realize 500 watts?

$$I_{primary} \times E_{primary} = P_{primary}$$

or

$$I_{primary} \times 100\,V = 500\,W$$

and

$$I_{primary} = 5 \text{ amps.}$$

To prove this:

$$\text{POWER INPUT TO TRANSFORMER} = \text{POWER OUTPUT}$$

Then:

$$I_{primary} \times E_{primary} = I_{secondary} \times E_{secondary}$$

or

$$5\,A \times 100\,V = 1\,A \times 500\,V$$

and

$$500\,W = 500\,W$$

POWER TRANSMISSION LINES

Why does the power company use transformers to increase the voltage of their high tension cross-country lines? To answer this we must look into the subject of POWER LOSS. When current flows through a conductor it must overcome the resistance of the conductor and HEAT is produced. The heat serves no useful purpose and is lost to the surrounding air. It would be only good business to keep LOSSES AT A MINIMUM and if current is causing the loss, then current must be kept at a very low value.

For an example, we will assume a small factory needs 10,000 watts of power at 100 volts. The current required would calculate as:

$$I \times 100 = 10,000 \quad \text{or} \quad I = 100\,A$$

Direct and Alternating Current

This is a lot of current. The wires would have to be large and the losses would be great to bring this current from a far away generator. To avoid this, the voltage is transformed (stepped up) to a very high voltage, sent across country and then transformed (stepped down) to the lower required voltage.

Continuing the example, assume the voltage is increased to 200,000 volts. With this voltage the current for the needed power is:

$$\text{I x 200,000} = 10,000\,\text{W} \qquad \text{I} = .05\,\text{A}$$

Have we not the same power?

$$100\,\text{A} \times 100\,\text{V} = 10,000\text{ watts}$$

or

$$.05\,\text{A} \times 200,000\,\text{V} = 10,000\text{ watts}$$

By reducing the current in the high tension lines the power loss is kept at a minimum. The complete system is drawn in Fig. 8-15. Notice the use of transformers to increase voltage and decrease current.

At the generator I x E = 1200 W or .1 A x 12,000 V = 1200 W
Cross-Country   I x E = 1200 W or .02 A x 60,000 V = 1200 W
City Lines   I x E = 1200 W or .1 A x 12,000 V = 1200 W
Home   I x E = 1200 W or 10 A x 120 V = 1200 W

Fig. 8-15. This system shows how voltage is stepped up so that current can be minimum. This reduces power loss in the transmission lines. Voltage and current values selected to illustrate text example. High tension lines may carry 500,000 volts.

Fig. 8-16. Power transmission lines such as these, extend 950 miles between the Columbia River and Southern California. (Pacific Gas and Electric Co.)

Fig. 8-17. A modern steam-electric turbine generator unit of 330,000 kilowatts capacity.

Fig. 8-16 shows power lines carrying electrical power cross-country at a voltage of 500,000 volts. Fig. 8-17 shows a large generator powered by a steam turbine which generates 330,000 KW. At the substation in Fig. 8-18 are circuit breakers capable of interrupting electricity flowing at 500,000 volts, within a fraction of a second. Fig. 8-19 shows a modern steam-electric generating plant.

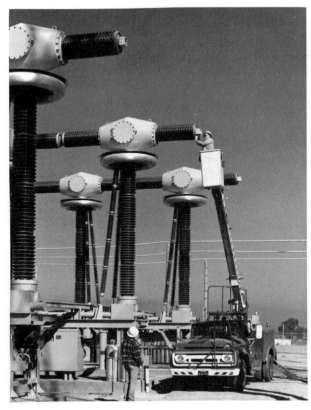

Fig. 8-18. These circuit breakers are capable of interrupting the flow of electricity being transmitted at 500,000 volts, within a fraction of a second.

Fig. 8-19. A typical modern steam-electric generating plant with a capacity of 2,113,000 kilowatts of electricity. The electricity is transmitted across the country at 230,000 and 500,000 volts. Natural gas fuel is used in the boilers.

RECTIFICATION

Power is supplied to our homes by local power companies at 110 and 220 volts at a frequency of 60 Hertz. But much of the electronic equipment like radios and TV's require direct current rather than alternating current. Also voltages other than 110 volts are needed. This introduces you to the means of changing ac to dc. This process is called RECTIFICATION. With the transformer and other components the circuit becomes the POWER SUPPLY.

DIODE RECTIFIERS, A ONE-WAY STREET

Like a one-way sign a DIODE RECTIFIER permits the flow of traffic in one direction only. A DIODE is a one direction device which permits current to flow easily in one direction, but offers an extremely high resistance to a current trying to flow in the opposite direction. The most common type used today is the solid state diode consisting of two types of silicon crystals. The symbol for the diode and a basic circuit which shows its operation is shown in Fig. 8-20.

Fig. 8-21. Assorted type of diode rectifiers and silicon controlled rectifiers. (International Rectifier Corp.)

$$I_{circuit} \times R_{load} = E_{output}$$

Assuming an ac source with a 10 volt effective voltage and a load resistance of 1000 Ω, the current would be .01 amperes. The output voltage would be:

$$I \times R = E \quad or \quad .01 \times 1000 = 10 \text{ volts}$$

EXPERIENCE 1. Secure from the school supplies, a diode and a 1000 Ω resistor and connect them, according to the schematic of Fig. 8-20, to a 6.3 volt a-c power source. Connect an oscilloscope first to the a-c source. Adjust the oscilloscope until four full a-c waves appear on the CRT (See page 159). Switch oscilloscope to ten volt range. The pattern should appear as in Fig. 8-22.

EXPERIENCE 2. Leaving oscilloscope adjustments the same, change the test leads to the output of the diode rectifier circuit. Compare to pattern in Fig. 8-22.

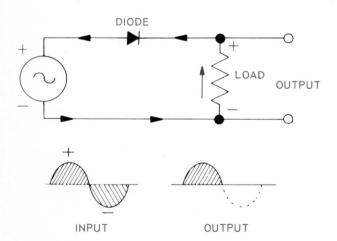

Fig. 8-20. The diode permits current to flow only in one direction. When voltage of source changes polarity the current drops to zero.

The photo, Fig. 8-21, shows typical diodes used in a rectifier circuit. These are made in many types, sizes and current ratings. The output voltage appears across the diode load and we know that voltage will be:

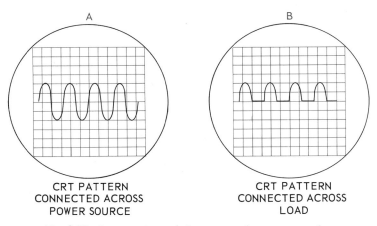

Fig. 8-22. A comparison of the input and output wave forms
of the diode rectifier circuit.

CONCLUSIONS: The output voltage no longer goes in a NEGATIVE direction. The output is, therefore, a PULSATING DIRECT CURRENT and is no longer an alternating current. The PEAK output voltage is approximately the same as the PEAK input voltage.

Fig. 8-23. During the first half cycle D₁ conducts and produces output voltage across R.

Fig. 8-24. During the second half cycle D₂ conducts and produces output voltage across R.

Since one-half of the a-c input wave is cut off and contributes nothing to the output, the circuit is called a HALF-WAVE RECTIFIER.

FULL-WAVE RECTIFIER

In order to secure full-wave rectification two diodes must be used. In full-wave rectification both half cycles of the a-c input will produce a useful output. However, the voltage output will be only one-half as much unless a more elaborate bridge rectifier with four diodes is used.

EXPERIENCE 3. A full-wave rectifier should be constructed by following the schematic diagram of Fig. 8-23. The transformer is a step-down to 12.6 volts for the secon-

dary. The secondary also has a center tap. In Fig. 8-23 the top of the secondary is positive and diode D₁ conducts producing an output voltage across the load. In Fig. 8-24 the bottom of the secondary is positive and D₂ conducts and produces output voltage across R. Therefore there is output during both halves of the a-c input wave. The voltage output is approximately 6.3 volts since it is taken from only one-half of the secondary winding.

Connect the oscilloscope across the secondary winding and adjust to display four complete wave forms (See page 159). Now connect the oscilloscope across R and adjust to display 8 half waves. Waves should appear as in Fig. 8-25.

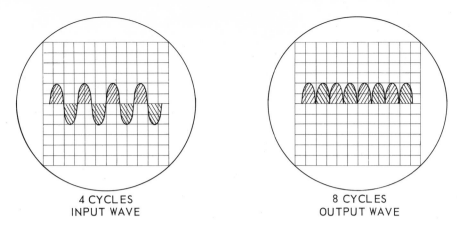

4 CYCLES
INPUT WAVE

8 CYCLES
OUTPUT WAVE

Fig. 8-25. Input and output waves of the full-wave rectifier circuit. The output is 2 times the frequency of the input.

CONCLUSION: The output of the half-wave rectifier has a frequency of twice the input wave and it is a pulsating d-c current.

FILTERING

Observe the roughness of both the half-wave and full-wave rectifier waves. Even though they are dc, they nevertheless are varying between zero and peak value. A constant value of d-c voltage is desired.

Fig. 8-26 shows a complete power supply consisting of a transformer and a rectifier followed by a filter. The action of the capacitors is important. At peak pulses the capacitor charges to peak voltage. When the pulse from the rectifier drops to zero, the capacitors discharge into the load. The filtering action tends to hold the output across the load at a constant value. The pulsations have been removed. THE INDUCTOR OR CHOKE OPPOSES ANY CHANGE IN CURRENT. As a result the output appears reasonably even and constant with very little RIPPLE or variation. Refer to waves in Fig. 8-27 and note filtering action.

OSCILLOSCOPE WAVE FORMS

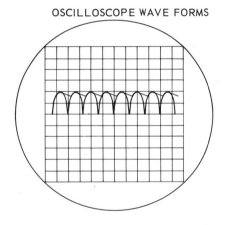

Fig. 8-27. Output wave form before (in black) and after filtering (in color).

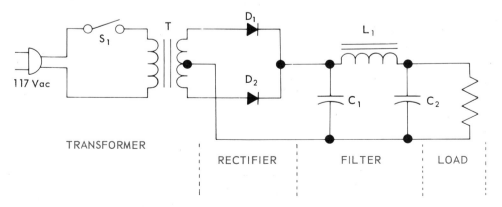

117 Vac

TRANSFORMER   RECTIFIER   FILTER   LOAD

Fig. 8-26. The complete power supply with FILTER.

CONDENSATIONS

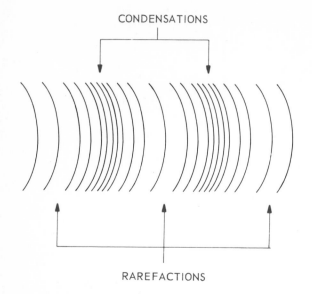

RAREFACTIONS

Fig. 8-28. Sound waves are condensations and rarefactions of air. Your ears respond to this sound wave.

AUDIO WAVE TO SOUND WAVE

An extremely important principle in electronics is the changing of an electrical wave, called the SIGNAL, to a sound wave which can be heard by the human ear. Your ears respond to sound waves. These waves come to your ear by vibrations of the air. The physicist describes a sound wave as alternate "condensations and rarefactions" of the air. To understand these words refer to Fig. 8-28. An electronic device which changes electrical waves into sound waves is a SPEAKER. Electrical waves can cause a diaphragm of paper or fiber to vibrate and produce sound waves. To understand this action study Fig. 8-29. Here a small coil of wire, called a VOICE COIL, is suspended in a permanent magnetic field. Attached to the voice coil is a paper cone. A varying audio signal current connected to the voice coil will cause the coil to move outward and inward at the frequency of the audio wave. The cone makes the sound wave for you to hear. This is the principle of the permanent magnet or PM SPEAKER.

Human hearing has its limitations. The range is generally considered between 20

Fig. 8-29. In the PM speaker the cone is caused to move in and out at the frequency of the input signal by the interaction between a permanent magnetic field and the electromagnetic field of the voice coil.

Fig. 8-30. A high fidelity 3-way speaker. (Universal Sound)

The following labels appear in the figure:

TERMINAL POST
TENSILE LEAD
CERAMIC RING MAGNET
CENTERING SPIDER
CERAMIC RING MAGNET
VOICE COIL AND FORMER IN VOICE COIL GAP
VOICE COIL AND FORMER
CONOIDAL LOADING RING
FOAM DAMPING PAD
SPHERICAL DIFFRACTOR
TWEETER DOME DIAPHRAGHM
POLE PIECE
POLE PIECE
FOAM DUST CAP
MASS LOADING RING
MAGNET RETURN CIRCUIT
VOICE COIL GAP
MID-RANGE DIFFUSICONE ELEMENT
CROSSOVER CAPACITOR
MAGNET RETURN CIRCUIT
WOOFER CONE
RIM GASKET
TREATED SURROUND
DIE CAST BASKET

and 20,000 Hertz or cycles per second. Low tones have low frequencies; higher pitch tones have higher frequencies. Typical examples are:

Man's voice - average. . . . . . .125-130 Hz
Woman's voice - average. . . . .250-260 Hz
Violin, up to approximately . . . . . 4,000 Hz

Middle A on piano . . . . 440 Hz
Meadow Grasshopper singing. . . 14,200 Hz
Baby Robin tweeting. . . . . 15,000 Hz

The ability of a speaker to produce sounds of different frequencies depends on its size, construction and quality. A 3-way speaker is pictured in Fig. 8-30. This contains a WOOFER for low frequencies, a TWEETER for the highs and an INTERMEDIATE SPEAKER for the in-between ranges. It is a HIGH FIDELITY SPEAKER.

FORWARD STEPS IN UNDERSTANDING ELECTRICITY-ELECTRONICS

1. A generator converts mechanical energy into electrical energy.
2. To induce a voltage there must be a conductor, a magnetic field, and relative motion between them.
3. Alternating current changes polarity and direction.
4. Frequency is measured in cycles per second in units called HERTZ.
5. An alternator generates a-c current by rotating a coil in a magnetic field or by rotating a magnetic field within the stator coils.
6. A commutator is a mechanical switch which changes the output of a generator to pulsating dc.
7. Generator output is controlled by controlling the current which produces the magnetic field.
8. A transformer transfers electrical energy from one circuit to another by means of magnetic fields.
9. The input coil of a transformer is the Primary.
10. The output coil of a transformer is the Secondary.
11. Voltage output of a transformer depends on the turns ratio.
12. In a transformer "power in" equals "power out" assuming no losses.
13. A transformer can be step-up or step-down or 1 to 1, depending on turns ratio.
14. Transformers are used by power companies to step up voltage for power transmission to decrease line losses.
15. A DIODE conducts electricity in only one direction.
16. Rectification means the changing of ac to dc.
17. A half-wave rectifier uses one-half the input cycle to produce a useful output.
18. A full-wave rectifier uses the complete input cycle to produce a useful output.
19. A rectifier produces a pulsating d-c voltage and a filter is required to remove the pulsations.
20. Capacitors tend to keep the output voltage at a constant value.
21. Chokes or inductors tend to keep the output current at a constant value.
22. A PM speaker uses a permanent magnet.
23. A speaker converts a-c waves at audio frequency into sound waves, by interaction of magnetic fields.
24. The signal to a speaker is connected to the voice coil.

TEST YOUR KNOWLEDGE - UNIT 8

1. A transformer has a 500 turn primary. How many secondary turns are required to give an output of:
   1000V, 25V, 400V, 18V, 6V
   NOTE: Imput voltage is 100 Vac.
2. What would happen if a transformer primary were connected to a d-c voltage?
3. How can an automotive alternator be used to charge an auto storage battery?
4. Draw the circuit of a power supply which could be used for a 9 volt transistor radio.
5. A factory uses 10,000 watts of power. A 500,000 volt transmission line brings power from generator to distribution station, to factory. What current must flow in the 500 KV line to supply the factory?
6. Pay a visit to your local power company and find out for your area:
   1. Generator Source Voltage.
   2. High Tension Line Voltage.
   3. Distribution Line Voltage.
   4. City Branch Line Voltage.
7. If there is a power line near your home answer these questions: Is there a transformer on it? What is the purpose of the transformer? How many homes on your street are served from one transformer?
8. Which is easier to filter, the output of a half-wave rectifier or a full-wave rectifier? Why?
9. A fun "shocker" circuit is sketched in Fig. 8-31. This uses a 1.5 volt D cell. It will produce a surprising but not dangerous shock. Build the circuit and explain its action. Why do you get no

PUSH BUTTON

1.5V
FLASHLIGHT
CELL

METAL
HANDLES

SPEAKER TRANSFORMER
FROM OLD RADIO

Fig. 8-31. Hold the metal handles and push the button and release it. You'll be surprised.

shock when you hold the push button down?

10. A step-up transformer has a 100 volt input and a 500 volt output voltage. Which winding, the primary or secondary, will use the larger size wire for its coil?

THREE-WIRE CABLE

Receptacles, wires and appliances which meet modern standards of safety have a third or ground wire. A 3-prong plug is used. This is for your protection. Be certain that tools you use, such as drills, sanders and other machines, have their cases or frames connected to the ground.

SHARP METAL AND CORNERS

Small but painful cuts and bruises can happen when constructing metal chassis and cabinet enclosures. File the sharpness from the edges. Remove burrs made by drills and reamers. Always get FIRST AID for minor cuts.

A typical Control Memory Board used in a computer. Illustrates the density of integrated circuitry used in the system. This board contains 64 words of six bits each.
(Honeywell Electronic Data Processing)

Unit 9

ELECTRONIC COMPONENTS IN ACTION

In your studies of electronics you will want to build many interesting and useful circuits. It is a major purpose of this book to open the doors to exciting adventures in electronic science. You may wish to continue on to advanced studies and a rich and rewarding career in electronics.

The skeleton, the backbone, the plan of all electronic circuits is the SCHEMATIC DIAGRAM. Like the carpenter about to build a house, he must first secure the architectural plans to find the sizes of rooms and their arrangement. The electronic technician must secure his plan to find the size and arrangement of components like resistors, capacitors and inductors. The components and devices must be connected together to produce the desired effects. To learn to read and understand electronic components is a very important part of your studies. In Fig. 9-1 is drawn a schematic of a simple transistor radio. The components shown are made in a wide variety of sizes and values.

RESISTORS

You have already studied the effect of resistance in a circuit in Units 3 and 4. Resistors may be connected in series or in parallel. In either case, resistance is an opposing force to the flow of electricity. When a current flows through a resistor, energy is used and creates heat. In your electric range at home, resistance is used to produce heat for cooking. On the other

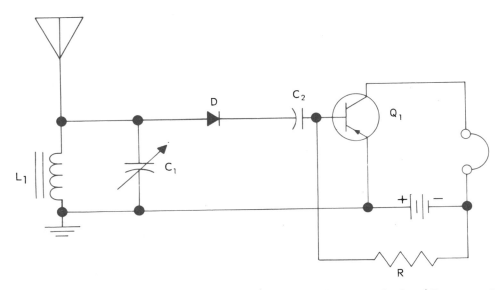

Fig. 9-1. A simple schematic diagram of a one transistor radio. Notice L, C and R components.

Fig. 9-2. Fixed carbon resistors. They are made in a wide variety of ohmic values and several wattage sizes.

hand, heat produced by current through resistance in your Hi-Fi amplifier could be destructive if provisions were not made to radiate this heat into the surrounding air.

Fig. 9-3. Wire wound power resistors.

The common type of resistor is the carbon resistor shown in Fig. 9-2. The symbol is also shown. These are made in a wide variety of values from only a fraction of an ohm up to millions of ohms. You will note a difference in physical size. The larger sizes have a greater radiation surface to dissipate the heat. These resistors can be purchased in 1/4 watt, 1/2 watt, 1 and 2 watt sizes.

For higher currents which produce more heat, a larger resistor is made by winding resistance wire on a ceramic form. These

are called POWER RESISTORS, Fig. 9-3. They are made in 25, 50, 100 watt sizes and even larger.

A variation of the power resistor has a slider on the body of the resistor which can

Fig. 9-4. Wire wound resistors with adjustable slider. (Ohmite)

be adjusted for some required resistance value. The adjustable power resistor and its symbol are shown in Fig. 9-4.

Fig. 9-5. Wire wound rheostat for variable resistance in power circuits. (Ohmite)

Variable resistors are illustrated with their symbols in Figs. 9-5 and 9-6. These may be made of carbon or resistance wire depending on power requirements. They are

Fig. 9-6. Carbon potentiometer. These are made in many ohmic values. (Ohmite)

available in many values. They are called POTENTIOMETERS or RHEOSTATS depending on how they are used. Such a component is used for the Volume Control on your radio. In Fig. 9-7 potentiometers are used in a television circuit to adjust Height and Width of picture.

CAPACITORS IN ACTION

This new component in our studies (CAPACITOR) will require some thought. A capacitor consists of two conducting plates separated by an insulator or nonconducting

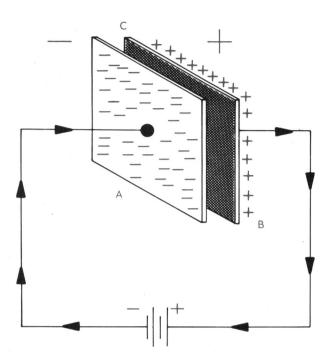

Fig. 9-8. Electrons flow from source to plate A. Electrons are repelled from plate B and are attracted to positive terminal of battery. Capacitor becomes charged.

material. There is no electrical path through a capacitor for direct current. Follow the circuit in Fig. 9-8. Electrons flow to plate

Fig. 9-7. Potentiometers are used on this television chassis for adjustments to picture height, linearity and width. These adjustments are made by the serviceman. They are usually found on the backside of a TV. (Centralab)

A. Because plate B is so close to plate A, the electrons are forced from plate B and are attracted to the positive terminal of the battery. The CAPACITOR becomes CHARGED at a polarity opposite to the battery. Current only flows in the circuit for a fraction of a second while C is being charged. The capacitor now is a source of power in its own right. It can be removed from the circuit and used to light a small lamp. HIGHLY CHARGED CAPACITORS CAN DELIVER SERIOUS SHOCKS AND BURNS, IF IMPROPERLY HANDLED.

this question: How much of a charge can a capacitor hold? The unit of measurement of CAPACITY is the FARAD (f) which means it will hold a specified number of electrons when one volt is applied. In electronics, smaller units for measuring capacity are used. They will be marked in microfarads (μ f) and picofarads (p f). A microfarad is 1/1,000,000 of a farad. A picofarad is 1/1,000,000 of 1/1,000,000 of a farad. Also and very important, a capacitor will have a WORKING VOLTAGE RATING, and it will be so marked. If this voltage is exceeded,

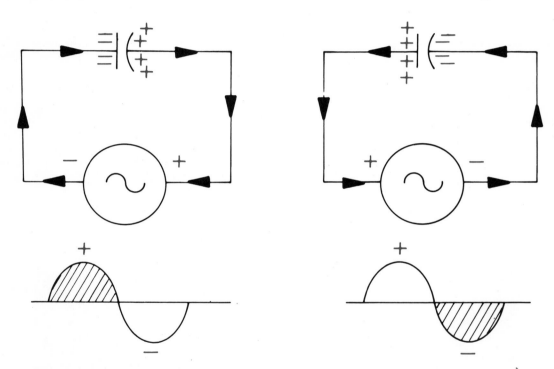

Fig. 9-9. The capacitor charges and discharges according to the a-c voltage source.

CAPACITORS AND AC

When connected to an alternating current source, a capacitor performs in a different manner. The capacitive circuit is redrawn, using a symbol, in Fig. 9-9. When the ac is on its positive half-cycle the capacitor charges to one polarity. When the ac is on its negative half-cycle the capacitor charges to the opposite polarity. Current is flowing in the circuit at all times; first in one direction and then the other. You may raise

the capacitor will short out and be destroyed.

EXPERIENCE 1. Connect a 50 μ f, 50 volt capacitor in series with a 6 volt lamp to your 6 volt d-c supply. See Fig. 9-10. Does the lamp glow?

CONCLUSION: A capacitor blocks dc and the lamp will not glow.

EXPERIENCE 2. Use the same circuit but connect it to a 6 volt a-c power source.

Fig. 9-10. A capacitor blocks dc but permits some ac to flow to light the lamp.

Does the lamp glow? Does it glow as brightly as it should?

CONCLUSION: Yes, the light does glow when connected to ac. There is an a-c current in the circuit due to the charge and discharge of C. But it does not glow brightly. The capacitor is not large enough. It has insufficient capacity. Not enough current is allowed to flow.

EXPERIENCE 3. Connect another 50 μf capacitor in parallel to C_1. Refer to circuit, Fig. 9-11. Now the lamp glows much brighter.

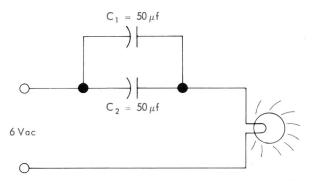

Fig. 9-11. A second capacitor C_2 is connected in parallel to C_1. The capacitance has doubled and the current has increased.

CONCLUSION: The capacitance of the circuit is now twice as much and the charging current causes the lamp to glow brighter.

Thinking about these last two experiences you might come to another conclusion. The single capacitor permitted some current to flow. Increasing the capacitance by adding C_2 permitted more current to flow. That sounds like resistance by another name. That is true. The opposition of a capacitor to the flow of alternating current is called REACTANCE. It, too, is measured in ohms, like resistance. Reactance in ohms depends on the size of the capacitor and the frequency of the alternating current.

TIME CONSTANTS

An extremely useful capacitive action is the time required to charge a capacitor. Look over the circuit in Fig. 9-12. The capacitor is connected in series with a resistor to the 15 volt d-c power source. When the circuit is turned on, the capacitor cannot charge in an instant. The resistor limits the electron flow. Some time must pass before C becomes charged to about full voltage. The TIME CONSTANT of the circuit is the time required for C to charge to about 63 percent of its final value. This interval of time can be computed by the formula:

Fig. 9-12. When power is turned on, C charges slowly as indicated on the voltmeter.

Fig. 9-13. A laboratory setup for demonstrating the charge and discharge of a capacitor.

R in ohms x C in farads = T in seconds

At the end of 5 time constant periods the capacitor is assumed to be charged.

EXPERIENCE 1. A circuit of a 10 μf capacitor and a 1 megohm (1,000,000 Ω) resistor is connected in series to the 15 volt d-c power source, in Fig. 9-12. To watch C charge, a voltmeter (VTVM) is connected across its terminals. Turn on the power and observe the movement of the meter.

EXPERIENCE 2. After C is charged in EXPERIENCE 1, disconnect the two leads to the power source and connect the two leads together. C will now slowly discharge as indicated by the meter.

CONCLUSIONS: The rate at which a capacitor charges and discharges depends upon the resistance and capacitance of the circuit. A laboratory setup for demonstrating the charge and discharge of a capacitor is shown in Fig. 9-13.

Fig. 9-14. Variable capacitors are manufactured in many sizes and types to fill the requirements of electronic circuits. (J. W. Miller Co., and Hammarlund Mfg. Co.)

THE CAPACITOR FAMILY

A variable capacitor is illustrated in Fig. 9-14. Note the plates and how they mesh together when the rotor is turned. In the fully meshed position there is maximum capaci-

Fig. 9-15. A small adjustable *TRIMMER* capacitor. Used for only small changes in capacitance. These adjustments are made by a technician when your equipment needs service. (Centralab)

tance. The insulation between the plates is AIR. In Fig. 9-15 a small variable TRIMMER capacitor has mica insulation between its plates. It is adjusted by a screwdriver.

Fig. 9-16 pictures the common paper capacitor. The plates are tinfoil separated by waxed paper. It is rolled up in tubular form and coated with wax or plastic. These

Fig. 9-16. A molded paper capacitor.

are made in many values and voltage ratings. Get a paper capacitor from an old radio or TV. Cut it open with a knife and see how it is made.

Dozens of small ceramic disc capacitors as illustrated in Fig. 9-17 can be found in a TV set. These are used where small values of capacitance are required. A ceramic disc, which is the insulator, has silver deposited on each side for plates. Wire leads are attached to the plates and the whole device is coated with a protective material. Values of this type of capacitor are usually in picofarads.

Fig. 9-17. Ceramic disc capacitors. (Centralab)

When a large amount of capacitance is needed and there is limited space, an ELECTROLYTIC capacitor is useful. A tubular type appears in Fig. 9-18. These are formed by a strip of aluminum coated with oxide and a layer of paper or gauze soaked in a paste-like electrolyte. A metal contact plate completes the sandwich. It is then rolled up in tubular form like the paper capacitor. The electrolytic capacitors usually have values

Fig. 9-18. A tubular type electrolytic capacitor.

from 1 to 1000 or more microfarads and many working voltage ratings. It is important to remember that the ELECTROLYTIC CAPACITOR HAS A PLUS AND MINUS POLARITY. IT MUST BE INSTALLED CORRECTLY. Failure to do so will ruin the capacitor and your circuit may not work. Sometimes they spark and smoke.

A can type electrolytic capacitor may be seen in Fig. 9-19. These are mounted on the top of a chassis. Some cans contain three or four capacitors of different values.

Fig. 9-19. A can type electrolytic capacitor. It can be rigidly mounted on a chassis.

A popular electrolytic capacitor used in transistor circuits and on etched circuit boards is the tiny low voltage electrolytic capacitor displayed in Fig. 9-20.

Fig. 9-20. LITTL—LYTIC electrolytic capacitors used in transistor circuits.

COILS AND INDUCTORS

In your studies of magnetism you have already come face-to-face with the solenoid coil and electromagnet. Coils are also used for many other purposes. Let's investigate the action of a coil in a circuit.

EXPERIENCE 1. A coil is connected in series with a 6 volt lamp to a 6 volt d-c power supply in Fig. 9-21. The lamp glows at full brightness.

Fig. 9-21. The coil and lamp are connected to a d-c source. The lamp burns brightly.

CONCLUSION: The coil seems to have little effect upon the flow of direct current. The only resistance in the circuit is the resistance of the wire in the coil.

EXPERIENCE 2. Connect the same coil and lamp to a 6 volt a-c power supply. The lamp now glows dimly. Refer to Fig. 9-22.

Fig. 9-22. The coil and lamp are connected to an a-c source. The lamp burns dimly.

CONCLUSION: The coil must oppose the flow of an alternating current. That is true. This opposition to ac is called INDUCTIVE REACTANCE and is measured in OHMS just like resistance.

EXPERIENCE 3. Continue EXPERIENCE 2, by placing a round iron core in the coil. See

Fig. 9-23. Now the lamp burns very dimly or not at all. Move the core in and out and note how the brightness of the light will change.

Fig. 9-23. An iron core is placed in the coil. The lamp is very dim or not burning at all.

CONCLUSION: By placing the core in the coil, the coil now MUST HAVE MORE REACTANCE. Much less current now flows.

COUNTERELECTROMOTIVE FORCE (CEMF)

You will recall that another name for voltage is electromotive force or EMF. Voltage causes a current to flow in a complete circuit. Counterelectromotive force or CEMF is a counter voltage developed which opposes the source voltage. How can that happen?

Fig. 9-24. The rising voltage and current causes an expanding magnetic field which cuts across the coil windings and induces a counter voltage of opposite polarity.

Study Fig. 9-24. A coil is connected to an alternating current supply. Consider the first half-cycle. As the applied voltage increases in the positive direction, the MAGNETIC FIELD also increases in strength and expands outward. The magnetic field expand-

ing outward cuts across the windings of the coil itself and INDUCES A VOLTAGE OF OPPOSITE POLARITY TO THE SOURCE VOLTAGE.

Now you should understand why the lamp glowed dimly when the coil is in series with it. Current in a circuit depends upon voltage. The current in this circuit is produced by the NET VOLTAGE or SOURCE VOLTAGE MINUS THE COUNTER VOLTAGE.

INDUCTIVE KICK

If a magnetic field is expanding and collapsing at a rate of 60 times per second, a certain CEMF will be developed. If the rate is speeded up to 1000 Hz, then a greater CEMF will be developed. This is like the generator. Increasing the revolutions per minute of the generator increases the rate at which the conductors cut across the magnetic field. Therefore, a greater output.

EXPERIENCE 4. The circuit in Fig. 9-25 should be made. Note that the coil is energized when you push the button switch. When

Fig. 9-25. The rapid collapse of the magnetic field when the switch is opened produces a voltage high enough to light the NE2 glow lamp.

the switch is released, the magnetic field collapses and produces a voltage high enough to light the NE2 neon lamp. This lamp requires 65 volts to light it.

CONCLUSIONS: The amount of CEMF developed by a coil depends upon the rate at which the magnetic field collapses.

The foregoing studies and experiences have many applications in electronics and industry. When an inductive circuit is opened

INTERMEDIATE FREQUENCY TRANSFORMERS

ANTENNA AND OSCILLATOR COILS

RADIO FREQUENCY CHOKE

U Case   X Case   A Case   Z Case   A-7J Case

AUDIO TRANSFORMERS

Fig. 9-26. Typical inductors and transformers used in electronic circuits.

either by a switch or a relay, high voltages are induced which cause arcs across the switch contacts. This will burn up the switch contacts and cause interference to radio communications nearby.

HENRY (H)

When a coil or inductor appears in a circuit, it will be marked with a letter L. Inductance is measured in HENRYS. For your information, if a coil produces one volt of CEMF when the current through the coil changes at a rate of one ampere per second, it is said to have an inductance of ONE HENRY. Coils used in electronic work will have values in henrys, millihenrys (mH) and microhenrys (μH).

THE INDUCTOR FAMILY

Typical coils found in electronic circuits are illustrated and named in Fig. 9-26. Their names suggest their use. Notice that many of them are transformers. This is expected,

since a transformer is just another way of using coils. There are thousands of types and values in common usage. Some you will use in interesting circuits and experiments later in this text.

IMPEDANCE (Z)

Many circuits contain resistance, capacitance and inductance. Each has its own way of opposing the flow of alternating current. The total effect is called the IMPEDANCE (Z) of the circuit. It is also measured in ohms. You cannot add up these effects as you would in addition, since they work in different ways. Computation of impedance will be studied in your later courses in electronics.

SIGNS AND SYMBOLS

A complete listing of approved signs, symbols and abbreviations will be found on page 162 in the back of this book. Keep this in mind for ready reference.

FORWARD STEPS IN UNDERSTANDING ELECTRICITY-ELECTRONICS

1. A capacitor consists of two conducting plates separated by an insulator.
2. Capacitance is measured in farads, microfarads and picofarads.
3. The opposition to an alternating current by a capacitor is called CAPACITIVE REACTANCE. It is measured in OHMS.
4. A capacitor blocks direct current.
5. Resistors are rated in watts which means the ability to dissipate heat.
6. A variable resistor is called a potentiometer.
7. An inductor is a coil of wire.
8. Inductance is measured in henrys, millihenrys and microhenrys.
9. The opposition to an alternating current by an inductor is called INDUCTIVE REACTANCE.
10. Capacitance in a circuit is the property of a circuit to oppose a CHANGE IN VOLTAGE.
11. Inductance in a circuit is the property of a circuit to oppose a CHANGE IN CURRENT.
12. The time constant of an RC circuit is the resistance in ohms times capacitance in farads equals time in seconds.

Fig. 9-27. *The circuit and mold setup for making the NITE-LITE. Suspend prongs and circuit from support and pour resin. Allow at least one hour for resin to harden. NITE-LITE plugs into any 110 volt wall receptacle.*
(Bobbie Fikes, Alta Loma High School)

TEST YOUR KNOWLEDGE - UNIT 9

1. Build a NITE-LIGHT. Refer to Fig. 9-27. Solder the resistor and NE2 Lamp together. Remove the metal prongs from an old a-c plug and solder circuit to it. Arrange in a mold or wax paper cup according to your desires. The liquid

Fig. 9-28. *Ask the circuit a question. Press the button. Either the "yes" or "no" light will give you the answer. A 90 volt d-c supply can be used. Connect to terminals A and B.*

Fig. 9-29. The BLINKER circuit. The NE2 lamps flash back and forth.

resin consists of resin, catalyst, dye (any color) and acetone. Consult your teacher or the crafts shop. Why is the 22 K ohm resistor used in series with the NE2 Lamp?

2. Build a DECISION MAKER. Is your answer "YES" or "NO"? You will never know until you press the button. Let electronics make your decision. Build the circuit in Fig. 9-28. Note that a half-wave rectifier is included. If you have a 90 volt d-c source, connect it as shown and do not use rectifier circuit.

3. Build a BLINKER. This cute little circuit in Fig. 9-29 will blink back and

forth. You can build it into a toy animal for blinking eyes. The power source is a 90 volt battery. It will last for years.

4. DO NOTHING BOX. In the circuit, Fig. 9-30, the lamps flash back and forth in no particular pattern. Build one in a small metal box. Support the lamps in the top of the box by sticking them through rubber grommets.

5. Build a SEQUENCE FLASHER. See Fig. 9-31. This circuit flashes 1-2-3. An example of a sequence flasher is on the rear turn signals of some automobiles. You may wish to rig a set to a direction arrow. You can always put colored glass in front of the lights.

6. Build a CONTINUITY TESTER AND TEST LAMP. Construct the circuit in Fig. 9-32 first on a breadboard. After testing, you may build it in a permanent box. To show the uses of this device we will perform three operations.

   A. Measure Resistance of circuit. DO NOT PLUG CIRCUIT INTO A-C WALL OUTLET. Use your ohmmeter and measure the resistance between:

   A and B = Ω
   C and D = Ω
   F and D = Ω
   F and A = Ω
   B and C = Ω
   E and F = Ω
   F and C = Ω

Fig. 9-30. The circuit for the DO NOTHING BOX.

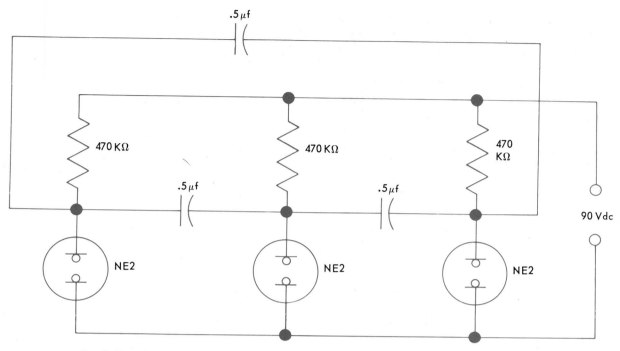

Fig. 9-31. This circuit flashes 1, 2, 3, and 1, 2, 3. Can you think of some clever uses for it?

Does the NE2 have any effect on your resistance measurement? Why?

B. Find the HOT side of an a-c power line. DO NOT PLUG THE CHECKER INTO AN A-C WALL OUTLET.

Connect the black test clip to the metal box or plate at a wall receptacle. Using the RED probe, stick it into first one slot in the wall receptacle and then in the other. One slot will light the NE2

lamp. That is the HOT side. It is very important to insert the plug of some equipment such as an ac-dc radio correctly into a wall plug. If incorrect, the chassis of the radio will be HOT or ALIVE. Accidentally touching this LIVE chassis, if your body is in contact with ground, will cause a severe shock and can kill you. Knowing the HOT side of a line is important.

C. MEASURE CONTINUITY between

Fig. 9-32. The circuit for the CONTINUITY CHECKER.

two points. Plug your CHECKER into the 110 volt wall outlet. Touch the RED probe to the BLACK clip. The light should light. Therefore, if the red probe is at one point and the black clip at another point and there is a connection or continuity between the points, the light will light. This information becomes useful in service and installation of electrical equipment.

RESISTORS ARE HOT COMPONENTS

Current flowing through resistance does work and produces HEAT. If your electronic equipment has been on for a while, some of the resistors get very hot. Watch out when working on the circuit. You can really burn your fingers.

CHARGED CAPACITORS DESTROY METERS

High voltages stored in capacitors may destroy a meter, especially ohmmeters, if measurements are being made. Always discharge capacitors to ground by touching an INSULATED SCREWDRIVER between the capacitor and the chassis ground, BEFORE servicing a circuit.

CAPACITORS STAY CHARGED

Many capacitors remain charged after the equipment is "turned off." Some retain a voltage which could knock you down. Always discharge capacitors by shorting the ends together with a jumper or an insulated screwdriver, BEFORE working on the circuit.

Unit 10

VACUUM TUBES
AND TRANSISTORS

We are truly indebted to the genius of a great American scientist, Thomas Edison.

Edison discovered early in his experimentation that an electric current flowing through a resistance wire would produce heat. How hot could he get the wire? Red hot? White hot? Hot enough to produce a glowing light? The problem was the discovery of some wire which would not burn up if hot enough to produce light. A really hot wire would oxidize with the surrounding air and destroy itself. The first successful incandescent lamp was made with a thread from Mrs. Edison's sewing basket coated with carbon and placed in a vacuum bottle. It burned for 40 hours. Lamps of today use tungsten wire for the filament and industry has found better ways of producing a vacuum.

WIZARD AT WORK

The story does not end there, for Edison continued his experimentation with "the glowing lamp in a bottle." Little wonder he was called the "Wizard of Menlo Park." Edison wondered if an electric current would flow through a vacuum. His original experiment was set up similar to Fig. 10-1. A voltage, now called the A supply, was connected to heat the light filament. A second voltage, now known as the B supply, was connected to make an independent circuit between the filament and a PLATE of conducting metal in the lamp.

Fig. 10-1. Edison's experiment which demonstrates an electric current can flow through a vacuum.

ONE WAY STREET

In Edison's first experiment, he placed the B battery so its NEGATIVE terminal connected to the plate. Nothing happened. When the B battery was turned around so its positive terminal connected to the plate, a CURRENT THROUGH THE TUBE was indicated on the meter.

Edison had produced a DIODE, (A ONE DIRECTION ONLY VALVE). Current would flow from filament to plate but not from plate to filament. The British still call this device a VALVE. The theory of operation of this diode remained a mystery for some years. At that time electric current was not known as "electron flow." In truth, the ELECTRON THEORY was not proposed until 1900 by J. J. Thomson. The explanation of this phenomenon seems quite commonplace

SOLID STATE
DIODE

VACUUM DIODE
DIRECTLY HEATED

VACUUM DIODE
INDIRECTLY HEATED

DUODIODE
(2 IN 1 TUBE)

Fig. 10-2. Symbols for solid state and vacuum diodes.

today. When Edison heated the filament, electrons "boiled" out and formed a cloud of electrons around the filament. This is called the Edison Effect. When the plate was "positive" it attracted the electrons through the vacuum and caused current to flow. When the plate was negative, it repelled the electrons.

Diodes through the years have had many applications as rectifiers and radio detectors. The vacuum diode is rapidly becoming obsolete by the development of the SOLID STATE DIODE. You have already used the solid state diode in your studies on rectification. The symbols for the vacuum diode and the solid state diode may be compared in Fig. 10-2. A full-wave rectifier power supply is drawn schematically in Fig. 10-3. Compare this circuit to Fig. 8-26 in Unit 8.

BIRTH OF ELECTRONIC AGE

At the turn of the last century, one of America's great scientists, Dr. Lee De-Forest, continued experimentation with the diode valve tube, by inserting a screen of wire between the heated cathode (the cathode emits the electrons) and the plate of the tube. Its effect had such a tremendous impact on electronic science that DeForest is often referred to as the "father of modern radio." He named this screen the GRID.

Now to see what happened. The symbol for a tube with a CONTROL GRID is shown in Fig. 10-4. It is called a TRIODE. Note that the grid is close to the cathode and all electrons must pass through the grid on their way to the plate. If the grid is made slightly NEGATIVE, it repels some of the electrons trying to pass through. Therefore, the electrons collected by the plate become less. If the grid is made more NEGATIVE only a

Fig. 10-3. A circuit for a full-wave rectifier power supply using a duodiode vacuum rectifier.

Fig. 10-4. A three element tube called the **TRIODE**. The heater is not counted unless it is also the cathode.

few electrons find their way to the plate. The GRID can be made NEGATIVE enough to prevent ANY ELECTRONS from passing through to the PLATE. The tube is said to be CUT OFF. A graph in Fig. 10-5 illustrates this action. Notice that as the grid becomes more negative, the plate current decreases until it becomes ZERO.

Fig. 10-5. Graph shows decrease in plate current as grid is made more negative.

MUCH FOR LITTLE

The astounding fact in the operation of the TRIODE (3 element) TUBE is this: A VERY SMALL CHANGE IN GRID VOLTAGE WILL PRODUCE A REMARKABLE CHANGE IN PLATE CURRENT.

A load resistor is connected in the plate circuit in Fig. 10-6. The plate current

must flow through this resistor. A voltage will therefore appear across this resistor (E = I X R). Assume a load resistor of 25,000 ohms. One milliampere of current would produce a voltage across the load resistor of:

$$.001 \text{ AMP} \times 25,000 \, \Omega = 25 \text{ volts}$$

Referring back to the graph in Fig. 10-5 you will see that it requires ONLY ONE VOLT CHANGE IN GRID VOLTAGE TO PRODUCE A ONE MILLIAMPERE CHANGE IN PLATE CURRENT WHICH CAUSES A 25 VOLT CHANGE ACROSS THE RESISTOR. THIS IS AMPLIFICATION. In this example, the voltage amplification is 25 times.

Fig. 10-6. A small change in grid voltage produces a large change in output voltage. This is voltage amplification.

BIAS

In order to faithfully amplify a signal (a-c voltage) we will select some fixed value of grid voltage and call this our OPERATING POINT or Q point. By doing this, the grid voltage can be varied up and down as well as the current can be made to vary around a fixed point. This fixed voltage is called the BIAS VOLTAGE.

The circuit is drawn in Fig. 10-7. The fixed BIAS is selected as negative 4 volts. The current at this point is 6 mA. An a-c generator with a one volt peak output is connected in series with the fixed bias voltage. When the a-c generator is zero: grid voltage is -4 volts. When ac is +1 volt: grid voltage is -3 volts:

Fig. 10-7. A 2 volt peak-to-peak change at the grid produces a 50 volt peak-to-peak change at the plate.

$$(-4) + (+1) = -3$$

When ac is -1 volt: grid voltage is -5 volts:

$$(-4) + (-1) = -5$$

As the a-c generator causes the grid voltage to vary back and forth from -4 to -3 to -4 to -5 the current through the tube varies from:

$$6 \text{ mA to } 7 \text{ mA to } 6 \text{ mA to } 5 \text{ mA}$$

The voltage across the load resistor varies from:

$$150V \text{ to } 175V \text{ to } 150V \text{ to } 125V$$

We therefore have an output of an a-c wave which has a peak-to-peak value of 50 volts (25 to 75) as the result of an input voltage signal with a two volt peak-to-peak value (-3 to -5). The gain is 25 times.

WHO TURNED IT OVER?

The output voltage appears to be upside down compared to the input. This is always true. A vacuum tube INVERTS the input signal. You can reason this out. In Fig. 10-7, at the fixed operating point, there are 150 volts across R which leaves 50 volts of the total voltage to appear across the tube. At 7 mA of current, then 175 volts appear across R but only 25 volts across the tube. At 5 mA of current, only 125 volts appear across R leaving 75 volts across the tube. As a result the voltage at the plate measured from ground is opposite in polarity to the input signal.

TUBE AS A SWITCH

There are numerous opportunities for a fast acting switch in electronic circuitry. A

Fig. 10-8. A vacuum tube can be used as a switch for high power circuits.

vacuum tube can be used for this purpose. Considering again the graph in Fig. 10-5, at negative 12 volts on the grid the tube is cut off. No current flows. It is like an open switch with infinite resistance. On the other hand, with zero volts at the grid the tube is approximately in full conduction. Therefore, a small positive voltage on the grid could trigger the tube into conduction. The plate current could activate a relay and thereby switch very high power circuits, motors and machinery. A switching circuit is illustrated in schematic Fig. 10-8.

MODERN VACUUM TUBES

Several types of vacuum tubes are illustrated with their symbols and BASE DIAGRAMS in Figs. 10-9A, 10-9B and 10-9C. Some have special grids such as the SCREEN GRID and SUPPRESSOR GRID to improve their performance. Tubes serve a wide variety of application: Miniature tubes for radio frequency signals, large tubes to handle high power, some tubes to amplify voltage, others to amplify current. Read the description of these typical tubes in the illustrations.

1—Glass Envelope

2—Internal Shield

3—Plate

4—Grid No. 3 (Suppressor)

5—Grid No. 2 (Screen)

6—Grid No. 1 (Control Grid)

7—Cathode

8—Heater

9—Exhaust Tip

10—Getter

11—Spacer Shield Header

12—Insulating Spacer

13—Spacer Shield

14—Inter-Pin Shield

15—Glass Button-Stem Seal

16—Lead Wire

17—Base Pin

18—Glass-to-Metal Seal

Fig. 10-9A. The internal structure of a miniature tube.

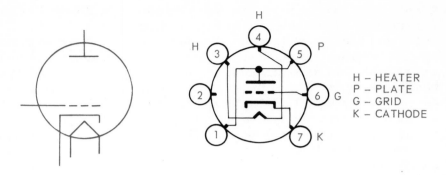

H – HEATER
P – PLATE
G – GRID
K – CATHODE

The 6C4 is a power **TRIODE** used in **FM** radios and other high frequency circuits.

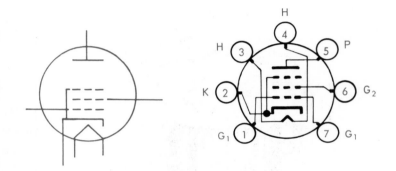

The 6AQ5 is a **BEAM POWER PENTODE** used in automobile radios and televisions sets. It has 3 grids.

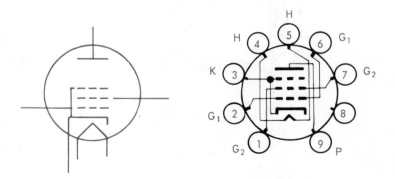

A 7868 power tube used in high fidelity amplifiers and radio receivers.

Fig. 10-9B. Several types of vacuum tubes with their symbols and base diagrams.

Fig. 10-9C. *Cutaway of NUVISTOR very high frequency tube.*

Labels on figure:
- CATHODE
- HEATER
- PLATE
- GRID
- METAL SHELL
- CERAMIC BASE WAFER
- INDEXING LUGS

THE TRANSISTOR

The invention of the transistor has revolutionized the electronics industry. They are tiny devices yet very rugged and dependable. They require "no heat for warm-up" like the vacuum tube. By using transistors we have many types of light, portable electronic

Fig. 10-10. *The transistor inventors (left to right) are Dr. William Shockley, now a professor of engineering at Stanford University; Dr. Walter Brattain, now retired after 37 years with Bell Telephone Laboratories; Dr. John Bardeen, now a professor of physics at the University of Illinois. All were engineer-scientists of the Bell Laboratories in 1948.*
(Bell Telephone Laboratories)

equipment. Space travel to the Moon and exotic and sophisticated computers were made possible by the transistor. The future of electronics is in the world of SOLID STATE DEVICES. The three scientists pictured in Fig. 10-10 received the Nobel Prize in Physics for their discovery of the transistor effect in 1948.

SOLID STATE CRYSTALS

Transistor and other solid state devices are mostly made of crystals of Germanium and Silicon, although other elements are used. These crystals in their pure form are very poor conductors of electricity. The crystals are changed by adding a very small quantity of an impurity which makes the crystal a semiconductor. It is neither a good nor a bad conductor.

ELECTRON CARRIERS

In one case an impurity is added which causes some free electrons to become available for conduction. This crystal is an N type crystal. Current flows through it in much the same manner as in conductors we have already discussed. An N crystal is displayed in Fig. 10-11. The current carriers are ELECTRONS.

ELECTRON MOVEMENT

N – CRYSTAL

Fig. 10-11. *An N type crystal conducts by electron movement like the typical conductor.*

HOLE CARRIERS

In the second case an impurity is added to the crystal which produces HOLES. Now

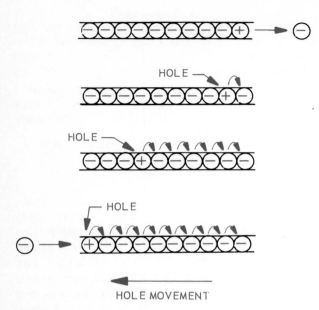

Fig. 10-12. *Conduction in P crystal by* **HOLE** *movement.*

a HOLE can be defined as a POSITIVE LO-CATION which has a strong attraction for an electron. These are called P type crys-tals. Current can also be conducted by HOLES. Study Fig. 10-12. An electron drawn to the positive power source leaves a va-cancy. The neighboring electron jumps over to the hole and leaves a vacancy. The next electron jumps into that vacancy and leaves a hole in turn. The HOLE, as you will ob-serve, moves from right to left, where it is finally filled by an electron from the nega-tive power source. In Fig. 10-13 conduction through a P crystal is illustrated. Hole movement is in the opposite direction to

Fig. 10-13. *In the P type crystal, the current carriers are* **HOLES.** *Electrons flow in external circuit.*

electron movement. Current in the outside circuit is always electron flow.

ELECTRONS MEET HOLES

A two crystal device is made by forming an N crystal and then changing the next sec-tion to a P crystal. The point of change is called the junction. This two element device is called a DIODE. In Unit 8, you studied and worked with this component. Remember it is a one-way street; a one-direction only conductor. Let's explore that. In Fig. 10-14 the diode of a P and N crystal are connected with the negative source to the P crystal and the positive source to the N crystal. Very little current flows.

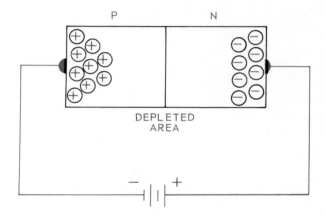

Fig. 10-14. *The crystal diode connected in a* **REVERSE BIAS** *direction. No major current flow.*

The electrons in the N crystal are at-tracted to the positive source.

The holes in the P crystal are attracted to the negative source.

Never the twain shall meet. Holes and electrons do not get a chance to meet at the junction. There is no major current flow in the circuit. This is stated as being RE-VERSE BIASED.

Now we will connect the diode in a FOR-WARD BIAS direction as in Fig. 10-15. The battery has been turned around; negative terminal to the N crystal and positive termi-nal to P crystal.

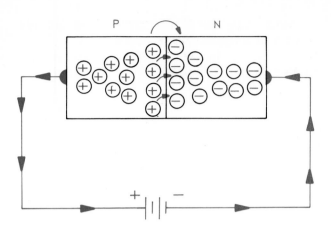

Fig. 10-15. The crystal diode connected in a FORWARD BIAS direction. Major current flows.

In this connection, electrons move from the source through the N crystal to the junction.

At the junction the holes in the P crystal are filled with electrons.

The electrons attracted by the positive power source create more holes in the P crystals.

The new holes move toward the junction to be refilled.

Current flows in the circuit. Electrons flow in the external circuit.

EXPERIENCE 1. Select a solid state diode from your electronics parts. Using the ohmmeter, measure the forward and reverse resistance of the device.

CONCLUSION: The diode must limit current to a very small amount in the reverse direction because its resistance is very high. The low resistance in the forward direction permits current to flow.

EXPERIENCE 2. Connect the diode, resistor, and milliammeter according to the schematic in Fig. 10-16. Measure the current with diode in forward bias connection. Remove diode from circuit and reconnect it in a reverse bias connection. Observe the current reading.

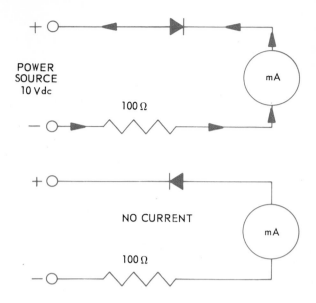

Fig. 10-16. Current flows only when diode is connected in forward bias. Resistor limits the current.

CONCLUSION: Current flows easily in the forward bias direction but very little flows in the reverse direction.

Diodes are manufactured in a wide range of sizes and current and power ratings. A typical TOP-HAT silicon diode is illustrated in Fig. 10-17.

Fig. 10-17. A typical silicon diode used in rectifier circuits. It is nicknamed the "top hat" because of its shape. (International Rectifier Corp.)

WHAT IS A TRANSISTOR?

A TRANSISTOR is a three element device made by forming two diodes back to back. The interesting results should encourage you to study this fascinating component in later courses. In Figs. 10-18 and 10-19 are block diagrams with a lot of information for an NPN transistor and again for the PNP. Two batteries are used, at this time, to assist you in understanding the action.

Fig. 10-18. *The NPN transistor connections.*

Notice that either the NPN or PNP transistor is connected so that the EMITTER-BASE diode is FORWARD BIASED. This is the low resistance direction.

Fig. 10-19. *The PNP transistor connections.*

Notice that the COLLECTOR-BASE diode is REVERSE BIASED in either type of transistor. This the very high resistance direction.

TRANSISTOR CURRENTS

Study the diagram in Fig. 10-20. Because the emitter-base circuit is forward bias, a current flows into the EMMITTER (E). The BASE (B) section is very thin and also electrons can quite easily cross the base-

Fig. 10-20. *Only a small percentage of the current flows in the base circuit.*

collector junction. As a result about 99 percent or more of the emitter current flows directly to the COLLECTOR (C). Only 1 percent flows in the BASE CIRCUIT. These simple additions are true:

$$I_E = I_B + I_C$$

$$I_C = I_E - I_B$$

CURRENT GAIN

The output current I_C must be LESS THAN the input current I_E. There can be no gain in current. In fact it must be less than one. The current gain is called ALPHA (α) and is less than one. It is equal to: $\dfrac{I_C}{I_E}$.

Assume an emitter current of 10 mA and a base current of .1 mA. Then:

$$I_C = I_E - I_B = 10\,mA - .1\,mA = 9.9\,mA$$

$$\text{Current gain} = \frac{I_C}{I_E} = \frac{9.9}{10} = .99 = \alpha$$

VOLTAGE GAIN

This is a different story. The resistance of the forward biased EB junction is low and the resistance of the CB junction is very high. About the same current flows through both junctions. Therefore the voltage across the EB junction is low and the voltage across the CB junction is high ($I \times R = V$). A circuit similar to a transistor shows these resistances and currents in Fig. 10-21. The voltage "in" and across junction

Fig. 10-21. I_E *flowing through low resistance.* V_E *is a small voltage.* I_C *(which is only 1 percent less than* I_E*) flows through high resistance* R_{CB} *and produces a high voltage.*

EB is $I_E \times R_{EB}$. The voltage "out" is $I_C \times R_{CB}$. Using our previous current figures and assuming $R_{EB} = 50\,\Omega$ and $R_{CB} = 500,000\,\Omega$, then the voltage gain (A_V) of the transistor is:

$$\frac{\text{OUTPUT}}{\text{INPUT}}\quad \frac{9.9\,mA \times 500,000\,\Omega}{10\,mA \times 50\,\Omega} = \frac{4950v}{.5v} = 9900$$

This gain is reduced by other resistors and components in a circuit but it still is very high.

NAME OF THE GAME

The previous discussion tells you how the TRANSISTOR got its name. TRANS stands for transfer and "sistor" for resistance. A transistor is a device which realizes its gain by a "transfer of resistance" or "transistor."

THE COMMON EMITTER CIRCUIT

By far the most popular way of connecting a transistor in a circuit is the COMMON EMITTER or CE method drawn schematically in Fig. 10-22. The input signal is applied between the base and emitter; the output is taken from the collector and emitter. The emitter is common to both circuits. A small voltage applied to the input produces a relatively large voltage across R_C in the output. R_C is the load resistor and works exactly the same as when used with the vacuum tube studied earlier in this chapter. The common emitter circuit also turns over or inverts the signal like a vacuum tube.

Fig. 10-22. *The* **COMMON-EMITTER** *circuit connection. The emitter is grounded and common to both the input and output circuits.*

COMMON EMITTER CURRENT GAIN

The ability of the common emitter circuit to amplify current may be understood

Fig. 10-23. The schematic diagram for the experimental amplifier.

by these examples. Assuming, as before, the emitter current of some transistor is 10 mA and the collector current is 9.9 mA; we can say that 99 percent of the total current flows in the collector circuit and only one percent in the base circuit. This percentage ratio stays more or less constant under a given set of circuit conditions. In other words, if the emitter current is increased to 20 mA then the collector current increases to 99 percent of 20 mA or 19.8 mA and the base currents is .2 mA. We can say then that a small change in BASE CURRENT results in a large change in COLLECTOR CURRENT. This relationship is called the BETA (β) of a transistor. It can be stated mathematically as:

$$\beta_{dc} = \frac{I_C}{I_B}$$

In our example:

$$\beta_{dc} = \frac{9.9 \text{ mA}}{.1 \text{ mA}} = 99$$

CASCADED TRANSISTORS

We talk about waterfalls on a mountain stream as CASCADES. One falls above another. To increase the gain or amplification

of a circuit, several transistors are coupled together in cascade fashion. The output of one feeds the input of the next stage.

FORWARD STEPS IN UNDERSTANDING ELECTRONS

1. The Edison Effect is the formation of a space cloud of electrons around a heated cathode.
2. A diode has a cathode and a plate. It conducts in only one direction.
3. A diode is used in rectification to change ac to dc.
4. A triode has a cathode, plate, and grid.
5. The grid in a triode controls the electron current flowing through the tube.
6. Amplification results when a small change in input voltage causes a large change in output voltage.
7. A triode amplifier inverts the signal.
8. Bias is the fixed voltage applied to the grid of a tube. It is usually negative.
9. A tube can act as a switch by switching grid voltage from conduction to cutoff.
10. A P crystal has HOLES for current carriers.
11. An N crystal has electrons for current carriers.

Fig. 10-24. A suggested breadboard layout for the amplifier.

12. A solid state diode conducts when forward biased.

13. A diode has high resistance in the reverse direction.

14. A transistor is a three element device and can be either NPN or PNP.

15. The current gain of a transistor connected as a common base circuit is called "alpha."

16. $I_E = I_C + I_B$, $I_C = I_E - I_B$,

$\propto = \dfrac{I_C}{I_E}$, $I_C = \propto I_E$

17. The common emitter is the most popular transistor connection. It produces both current gain and voltage gain.

18. Beta is the current gain of a common emitter circuit and indicates the ability of the circuit to amplify.

19. Transistors can be coupled in cascade for greater amplification.

TEST YOUR KNOWLEDGE - UNIT 10

1. Build a TWO TRANSISTOR AMPLIFIER CIRCUIT. Consult the diagram in Fig. 10-23 and the suggested layout in Fig. 10-24. Adjust voltage to 15 volts and turn the amplifier on.

A. Tap the microphone with your finger. The tap should be heard from the speaker. A very small voltage produced by the microphone is now amplified.

B. Place microphone near speaker. Does it squeal? This is caused by feedback. The noise from the speaker enters the microphone and is reamplified and around and around it goes.

C. Remove C_2 from the circuit. Does the amplifier work better or worse? The removal of C_2 reduces the gain of the amplifier by degeneration (feedback 180 deg. out-of-phase with input signal so it subtracts from input). The voltage at the emitter varies in such a manner as to oppose the input.

D. If there is a record player in the school shop, connect it directly to your amplifier. Does it amplify with reasonable fidelity?

E. Team up with a friend who has built a similar amplifier and build an intercom. You will need longer leads. Connect your microphone to his amplifier and his microphone to yours.

F. Connect an audio generator (see page 161) to the input of your amplifier. Set the frequency at about 400 Hz. Connect your oscilloscope across the input and adjust to display four waves about .2 volt from peak-to-peak. Connect the oscilloscope across the speaker terminals and observe the wave. Can you figure out the voltage gain? Is the output wave distorted? If so, reduce the A-F (audio-frequency) generator output until you have an undistorted wave.

2. Construct a TRANSISTOR SWITCH. To show the ability of a transistor to act as a switch, build the circuit in Fig. 10-25. Explain the transistor action.

Fig. 10-25. When S is closed, the lamp glows. The transistor is turned "on and off."

3. Build a TRANSISTOR RELAY. In order to switch higher currents and power circuits the relay circuit is constructed as in Fig. 10-26. The transistor circuit operates on 15 volts dc. The lamp circuit uses 6.3 volts ac.

4. What are the advantages of using transistors over vacuum tubes?

5. In an N crystal, the current carriers are _____.

6. In a P crystal, the current carriers are _____.

7. Two transistors are connected in cascade. One has a voltage gain of 50 and the other 60. What is the total voltage gain?

Fig. 10-26. A relay activated by a transistor switch. The relay contacts could control a high powered machine.

8. Build a SALINITY TESTER to test "how salty?" Water decreases its resistance when salt is added. In the circuit of Fig. 10-27 a glass jar filled with water is used. The test probes are pieces of bare copper wire. Add salt to the water a little at a time. Watch the current meter. The meter could read in percentage of salt rather than mA.

Explain the transistor action.

Fig. 10-27. The salinity of the water will be indicated on the meter.

Unit 11

ELECTRONICS IN THE COMMUNICATIONS INDUSTRY

The German scientist, Heinrich Hertz, discovered that an electric spark would set up a vibrating high frequency electric and magnetic field and these vibrations could be picked up at some distance by a coil of wire. Thus was born the possiblity of communications by vibrating electric waves.

Not until 1895 did the Italian scientist, Guglielmo Marconi, make the practical application of the Hertz waves. He commu-

these waves is limited by the mechanics of the machine. You can make a rotating armature spin just so fast or it will fall apart. Higher frequency waves are required for radio communications.

Electronic circuits are used to generate these waves. They are called OSCILLATORS. To oscillate means to swing back and forth or to vibrate. A playground swing is a good example of a swinging or oscillating move-

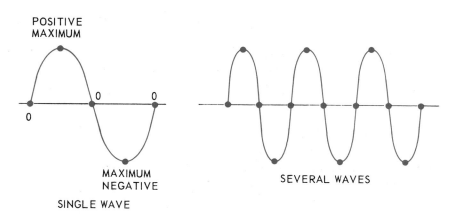

POSITIVE MAXIMUM

0

0   0

MAXIMUM NEGATIVE

SINGLE WAVE

SEVERAL WAVES

Fig. 11-1. Wave shows the varying voltage or current through cycle or alternation.

nicated with the Italian Navy twelve miles out to sea by vibrating waves. The first message sent overseas to America startled the world in 1901.

We have already studied the generation of alternating waves by means of the alternator, in Unit 8. However, the frequency of

ment. In electronics, we want a voltage or current to start at zero, rise to maximum in one direction or polarity, return to zero, then rise to maximum in the opposite direction or polarity and return to zero. We want this action to continue at some high frequency. The desired wave produced is similar to that wave illustrated in Fig. 11-1.

DESCRIPTION OF THE WAVE

So that we can talk about this wave and understand each other, let's put some names on the wave values and actions. See Fig. 11-2. The PEAK value of the wave is from zero to the peak value in one direction. The PEAK-TO-PEAK value is from maximum in one direction to maximum in the opposite direction. The FREQUENCY of the wave is the number of times the wave goes through its cycle per second. It is measured in Hertz. The PERIOD of the wave is the time in seconds it requires for one cycle. For example, a 100 Hertz wave requires 1/100 second for one complete cycle or oscillation.

Fig. 11-2. Parts of a wave. The effective value of the wave compares the alternating current wave to a direct current value.

EFFECTIVE VALUE

We learned early in our studies that current in a circuit is caused by voltage or pressure. In the alternating voltage wave the voltage never remains at a set value. It is changing continuously. If such a generator is connected to a circuit, the current also will change continuously. So we must have a way of comparing an a-c value with a d-c value. Since current through resistance produces heat and does work, we will discover then, "What peak value of ac will do the same work as some fixed value of direct current?" This has been named the EFFECTIVE VALUE OF THE A-C WAVE. It is the effective value of an a-c current or voltage which you usually measure with

Fig. 11-3. Follow the action of an energized LC circuit. A—S₁ open, S₂ open. B—S₁ closed, S₂ open C charges. C—S₁ open, S₂ closed C discharges through L and produces magnetic field. D—When C is discharged the magnetic field collapses and charges C in opposite polarity. E—C now discharges in opposite direction and produces magnetic field. F—When C is discharged the magnetic field collapses and charges C back to its original polarity. ONE OSCILLATION IS COMPLETE.

a meter. Its value may be found by simple arithmetic.

PEAK VALUE x .707 = EFFECTIVE VALUE

or

EFFECTIVE VALUE x 1.414 = PEAK VALUE

AN OSCILLATING CIRCUIT

Examine the circuit in Fig. 11-3. C is originally charged when S_1 is closed. When S_1 is opened and S_2 is closed, the capacitor discharges through the coil and produces a magnetic field. After C has discharged the collapsing field charges C in the opposite direction. The current oscillates back and forth in the circuit until the ENERGY originally stored in C from the power source is used up. Resistance in the circuit uses up a little energy with every current swing. Compare this to a playground swing. Each swing becomes less and less until the swing comes to rest. A current wave, gradually decreasing in height, is show in Fig. 11-4. It is a DAMPED WAVE, and becomes less and less with each oscillation.

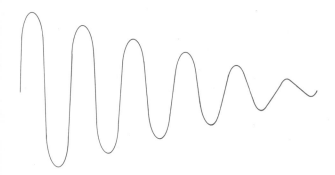

Fig. 11-4. The damped wave becomes less with each oscillation.

TUNED TANK CIRCUIT

The previous circuit of the capcitor and a coil is called a TANK circuit and the oscillating current is compared with water sloshing back and forth in a tank. The current action is sometimes referred to as FLY-WHEEL ACTION. The important point you must realize is: SOME VERY DEFINITE

PERIOD OF TIME MUST PASS DURING WHICH THE CIRCUIT COMPLETES ONE OSCILLATION. This time depends upon the VALUE OF L and C. For any set pair of values the circuit will oscillate at a particular frequency. To change the frequency of oscillation, it is only necessary to change either the value of C or L. This is called TUNING THE CIRCUIT.

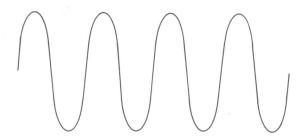

Fig. 11-5. Four cycles of a continuous wave. All peaks are equal.

CONTINUOUS WAVES

On the playground, the swing will continue to swing back and forth, if only a little push is added when the swing is at its backward position. The push replaces the energy used up by the friction of the swing. The same is true of the oscillating tank circuit. If a little pulse of electrical energy is added at the correct frequency, then the oscillator will continue to oscillate and produce a continuous wave (CW). Compare the CW in Fig. 11-5 to the damped wave of Fig. 11-4.

In electronic circuits a vacuum tube or a transistor is used as a fast acting switch to add energy to the tank circuit at exactly the right moment. This is the oscillator circuit. An amplifier feeds pulses of energy to the tank circuit. A small part of the energy in the tank circuit is fed back and reamplified to keep the tank circuit oscillating.

EXPERIENCE 1. Build an AUDIO OSCILLATOR. This circuit is so called because it produces a tone you can hear. Many young students build one for practicing the Morse Code such as used in Ham Radio

Fig. 11-6. The schematic diagram for the audio oscillator.

or Boy Scout work. The circuit diagram appears in Fig. 11-6 and the pictorial in Fig. 11-7. A key may be connected in the emitter lead to key the oscillator to sounds required for code practice. Turn the potentiometer and notice how it changes the pitch or frequency of the oscillator.

EXPERIENCE 2. Connect your oscilloscope across the speaker terminals. Adjust the oscilloscope for three stationary waves

about one inch high. Change the frequency of the oscillator and observe that the oscilloscope frequency must also be changed.

EXPERIENCE 3. Replace the $10\,\mu f$ capacitor with a $.1\,\mu f$ paper capacitor. Now the pitch is very high and sharp. Many electronic organs use circuits similar to these for producing the notes in the muscial scale. Such a circuit is used to test your hearing. Usually humans can hear up to

Fig. 11-7. Pictorial diagram of audio oscillator.

18 KHz. As you grow older, your hearing ability decreases. Dogs and animals sometimes hear sounds above 20 KHz.

RF OSCILLATORS

An oscillator which operates at frequencies above 300,000 Hz is generally classed as a radio frequency oscillator. Oscillators of this frequency and higher are attached to the radio transmitting antenna and the generated waves radiate into space. These are the waves you receive at home from the antenna on the roof of your house. These waves bring you the sounds of music and voice and television pictures.

RF AMPLIFIERS

The simple oscillator usually does not have enough power to generate a real strong wave and it would not radiate any great distance. To overcome this and assure you will receive the radio program, the oscillator output will be amplified by transistors and tubes to high power before it is coupled to the transmitting antenna.

CODE COMMUNICATIONS

A very reliable way of transmitting radio messages is by using the Morse Code. You, of course, must be able to understand the message. The code is extensively used by Amateur Radio and commercial radio stations.

Fig. 11-8. SOS written in code with radio waves. A short burst is a "dit" a longer burst a "dah."

For code communications the radio transmitter is turned on or off by a hand switch or key. A short burst of radiated energy represents a "dot" or "di" or "dit." A longer burst a "dash" or a "dah." For

example a short and a long burst, "dit dah" written (· —) means the letter A. In Fig. 11-8 the distress call for ships at sea "SOS" is written in code with the radio frequency waves above. The complete MORSE CODE used for this purpose may be found in Fig. 11-9. One must remember that this

A	· —	(di—dah)
B	— · · ·	(dah—di—di—dit)
C	— · — ·	(dah—di—dah—dit)
D	— · ·	(dah—di—dit)
E	·	(dit)
F	· · — ·	(di—di—dah—dit)
G	— — ·	(dah—dah—dit)
H	· · · ·	(di—di—di—dit)
I	· ·	(di—dit)
J	· — — —	(di—dah—dah—dah)
K	— · —	(dah—di—dah)
L	· — · ·	(di—dah—di—dit)
M	— —	(dah—dah)
N	— ·	(dah—dit)
O	— — —	(dah—dah—dah)
P	· — — ·	(di—dah—dah—dit)
Q	— — · —	(dah—dah—di—dah)
R	· — ·	(di—dah—dit)
S	· · ·	(di—di—dit)
T	—	(dah)
U	· · —	(di—di—dah)
V	· · · —	(di—di—di—dah)
W	· — —	(di—dah—dah)
X	— · · —	(dah—di—di—dah)
Y	— · — —	(dah—di—dah—dah)
Z	— — · ·	(dah—dah—di—dit)

1	· — — — —	(di—dah—dah—dah—dah)
2	· · — — —	(di—di—dah—dah—dah)
3	· · · — —	(di—di—di—dah—dah)
4	· · · · —	(di—di—di—di—dah)
5	· · · · ·	(di—di—di—di—dit)
6	— · · · ·	(dah—di—di—di—dit)
7	— — · · ·	(dah—dah—di—di—dit)
8	— — — · ·	(dah—dah—dah—di—dit)
9	— — — — ·	(dah—dah—dah—dah—dit)
0	— — — — —	(dah—dah—dah—dah—dah)

Period	· — · — · —	(di—dah—di—dah—di—dah)
Comma	— — · · — —	(dah—dah—di—di—dah—dah)
Question	· · — — · ·	(di—di—dah—dah—di—dit)

Fig. 11-9. The MORSE CODE.

code is sent over the air at a FIXED FREQUENCY. To receive the code, the RADIO RECEIVER MUST BE TUNED TO THE SAME FREQUENCY.

TRANSMITTING ANTENNA

RECEIVING ANTENNA

RADIO WAVE 980 KHz

CIRCUIT TUNED TO 980 KHz

CIRCUIT TUNED TO 980 KHz

Fig. 11-10. The radio receiver is tuned to the same frequency as the transmitted signal.

TUNING A STATION

You may wish to review our earlier discussion on tuned circuits. In Fig. 11-10 a typical diagram of a radio tuning circuit is drawn. In this case the capacitor is a variable capacitor controlled by a knob on the front panel of your radio. By varying the capacitance, the circuit will oscillate at any frequency between 550 KHz and 1600 KHz. If a radio station is broadcasting at 980 KHz, it is only necessary to tune the receiver to 980 KHz to hear the station. All radios including FM and television operate on this same principle, but use different frequencies and therefore different values of L and C in their tuned circuits.

Consider again the swinging playground swing. By pushing it at its most backward swing at the right time, you caused the swing to continue. Suppose you pushed the swing forward as it was moving backwards. Then you would oppose the swinging motion and the swing would quickly stop. The swing must be pushed at the same frequency or multiples of that frequency to keep the swing going. The receiver tuned circuit must be adjusted by varying the capacitance or inductance so that it will swing or oscillate when energy is added from the transmitted radio wave.

MODULATION

You would not enjoy a radio program sent by the Morse Code. In fact, you would need a lot of practice to even understand it. In order to send music and voice over the radio, something must be done. This is accomplished by MODULATION. Instead of transmitting a constant amplitude radio frequency wave, we must add circuits to the transmitter which will make the radio wave amplitude vary to match the frequency of voice or music. This is classed as AMPLITUDE MODULATION or just AM. Your broadcast band radio receives AM waves. To get a better picture of such a wave, study Fig. 11-11. Here a 400 Hz sound wave causes the radio frequency wave to change its height or amplitude at the rate of 400 Hz. This modulated radio wave now carries a 400 cycle tone as it is broadcasted into space. Many audio tones and speech can be used which result in a very complex modulated wave.

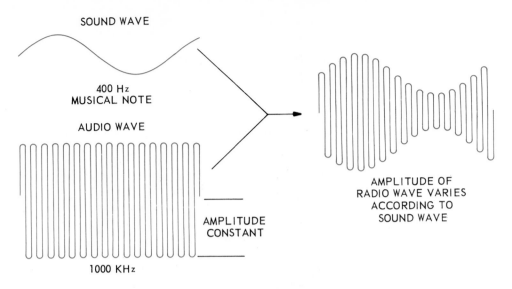

SOUND WAVE

400 Hz
MUSICAL NOTE

AUDIO WAVE

AMPLITUDE
CONSTANT

1000 KHz

AMPLITUDE OF
RADIO WAVE VARIES
ACCORDING TO
SOUND WAVE

Fig. 11-11. The result of changing the amplitude of a radio wave by a sound wave is Amplitude Modulation.

EXPERIENCE 1. Connect your oscilloscope to the output of a radio frequency generator. Set generator frequency at about 1000 KHz to produce a green band across the oscilloscope face about one inch in amplitude. Set the oscilloscope frequency near 400 Hz. Turn on the modulation in the RF generator and observe the modulated pattern. Slight adjustment of oscilloscope frequency may be necessary. Does it appear similar to FIG. 11-11?

EXPERIENCE 2. Without changing the oscilloscope controls, connect it across the speaker of any AM radio while a radio program is being received. Now you will see the complex pattern resulting from tones and sounds of many frequencies.

DETECTION

If an AM wave were tuned-in and connected directly to a speaker or earphones, you would hear nothing. Looking at the modulated wave again, you will see that for every positive peak there is an equal negative peak and the average value of the wave is zero. In order to detect the variation in amplitude of the wave which is the music or speech, one-half of the wave must be removed by rectification. In this application, it is called detection. A diode is used in Fig. 11-12 and both input modulated wave and the demodulated or detected output are illustrated. Only the positive half of the modulated wave causes an output since the diode cuts off the negative half of the cycle.

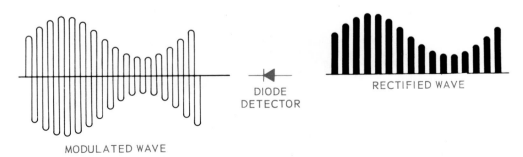

DIODE
DETECTOR

RECTIFIED WAVE

MODULATED WAVE

Fig. 11-12. After rectification, only the positive cycles of the wave remain. When modulated wave is negative the diode does not conduct.

PEAK VALUE

AVERAGE VALUE

THE DETECTED AUDIO WAVE

Fig. 11-13. The average value of the detected RF wave is the original sound wave.

Note: You could get a negative output by reversing the diode. This is sometimes done.

RECOVER THE MUSIC

In Fig. 11-13 the detected radio wave is again shown. Does it not consist of pulses of positive voltage of various amplitudes? It surely does. Since any cycle is only at its peak for a fraction of a second and continuously at some value between peak and zero at all other times, it must have an AVERAGE VALUE. This value is pointed out in the illustration. Observe that this VARYING AVERAGE VALUE wave has the same appearance as the original sound wave used to modulate the transmitter in Fig. 11-11. It can now be amplified and used to drive a speaker. The radio frequency wave remaining can be filtered out and bypassed to ground.

THE COMMUNICATIONS SYSTEM

A simplified BLOCK DIAGRAM of the transmitter and receiver is shown in Fig. 11-14. This method of representing circuits

is widely used in industry so that the technician can observe the total system and the individual units in the system.

FM RADIO

AM radio has its limitations when high fidelity music is broadcasted. It is also sensitive to noise and interference. In the frequency modulated or FM radio system the radio frequency wave is kept at a constant amplitude. The modulating sound wave causes the radio wave to change frequency at the same rate as the audio wave. The strength or loudness of the audio wave determines the amount of frequency change. Each of these conditions is described in the illustrations, Figs. 11-15 and 11-16. The FM broadcast band extends from 88 MHz to 108 MHz. In the example, the CENTER FREQUENCY of 100 MHz is used.

FM DETECTION

Although the circuits used to detect frequency variation are covered in more ad-

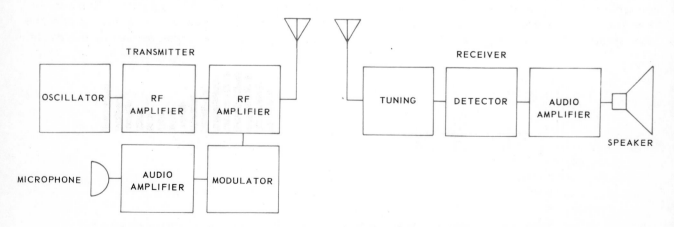

Fig. 11-14. The block diagram of a simplified radio transmitter and receiver.

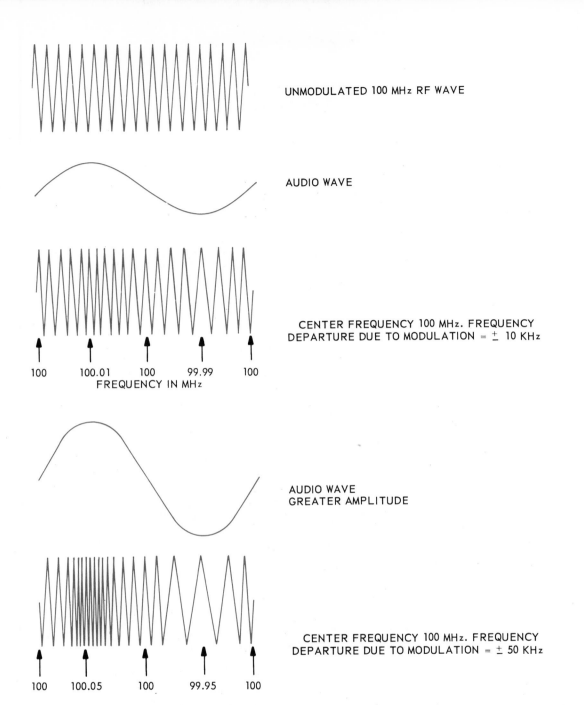

UNMODULATED 100 MHz RF WAVE

AUDIO WAVE

CENTER FREQUENCY 100 MHz. FREQUENCY
DEPARTURE DUE TO MODULATION = ± 10 KHz

100   100.01   100   99.99   100
FREQUENCY IN MHz

AUDIO WAVE
GREATER AMPLITUDE

CENTER FREQUENCY 100 MHz. FREQUENCY
DEPARTURE DUE TO MODULATION = ± 50 KHz

100   100.05   100   99.95   100

Fig. 11-15. The strength or amplitude of the sound wave de-termines the amount of departure from center frequency.

vanced textbooks, you should realize that this medthod of sending information over radio waves is distinctly different. A detector for AM must be sensitive to amplitude changes in a signal; a detector for FM must be sensitive to frequency changes. Each method must recover the sound waves used originally at the transmitter for modulation.

FM MULTIPLEXING

A tremendous advantage of FM radio is the capability of sending more than one radio program over the same frequency carrier wave. The familiar example is STEREO-PHONIC FM where the LEFT channel is separated from the RIGHT channel. At the

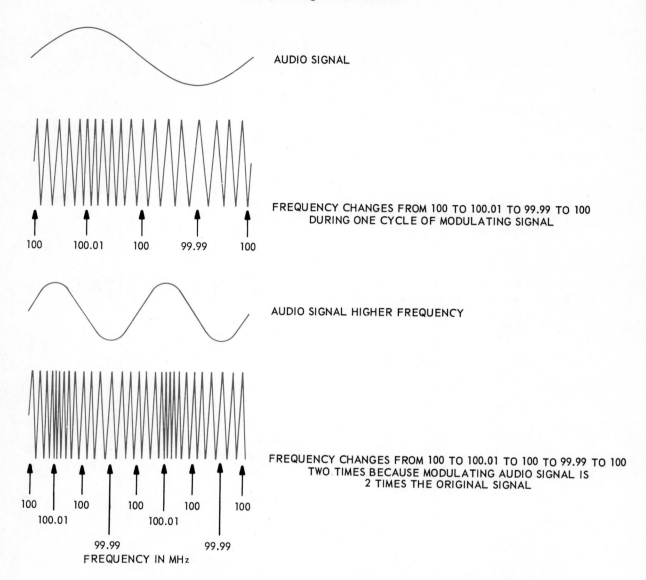

AUDIO SIGNAL

FREQUENCY CHANGES FROM 100 TO 100.01 TO 99.99 TO 100
DURING ONE CYCLE OF MODULATING SIGNAL

100   100.01   100   99.99   100

AUDIO SIGNAL HIGHER FREQUENCY

FREQUENCY CHANGES FROM 100 TO 100.01 TO 100 TO 99.99 TO 100
TWO TIMES BECAUSE MODULATING AUDIO SIGNAL IS
2 TIMES THE ORIGINAL SIGNAL

100   100   100   100   100
   100.01   100.01
        99.99   99.99
FREQUENCY IN MHz

Fig. 11-16. The **RATE OF FREQUENCY** *change depends
upon the frequency of the audio modulating signal.*

receiver they are detected and amplified separately and connected to LEFT and RIGHT speakers. This produces a musical front which gives the same effect as listening to a symphonic orchestra in a theatre, if this kind of music is being broadcasted.

TELEVISION

What is a picture? Examine a photograph in a newspaper or magazine with a magnifying glass. You will see that the picture is made up of thousands of small dots or areas varying from white to black (or color) and all shades in between. These are called picture elements. The arrangement of these elements makes the picture.

The video camera at the studio looks at the scene through a lens similar to a camera. Instead of film, the video camera uses a photo sensitive plate of many photo cells. Each element focused onto the plate produces a voltage corresponding to the darkness or lightness of the picture element. An electron beam in the video camera scans (scan means to look over in detail from point to point) the photo sensitive plate. The electron beam moves from left to right and downward in the same fashion as you are reading this printed page. It reads very fast however. In fact the picture contains 525 lines and the beam reads these lines 30 times per second. It therefore produces 30 complete pictures per second. The lightness or darkness of the picture element is changed into an electrical signal. The darker elements produce larger signals.

Fig. 11-17 shows a checker board and the signal developed in the camera as the electron beam scans two lines of the picture. The signal is used to modulate an AM transmitter in the same manner as for radio broadcasting.

PICTURE TUBES

A simplified drawing of a television picture tube is shown in Fig. 11-18. It works similar to the vacuum tube. Note the HEATER and the CATHODE which pro-

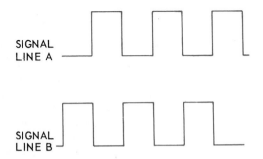

Fig. 11-17. The camera signal developed as electron beam scans two lines of a checkerboard.

Fig. 11-18. A simplified diagram of a picture tube. See text.

duces the space cloud of electrons. GRID No. 1 is the control grid and the number of electrons passing through it depends upon its negative voltage. GRIDS 2, 3, 4 and 5 are used for acceleration of the electrons toward the screen and for focusing the electron beam to make a tiny spot, like focusing a flashlight. When the beam hits the florescent screen, it causes the screen to glow at the degree of whiteness depending on the intensity of the beam. This is controlled by GRID No. 1. The electrons are collected by the high voltage anode to complete the circuit.

SCANNING

Around the neck of the picture tube in Fig. 11-18 are magnetic coils called the DEFLECTION YOKE. By varying the current through these coils, the beam can be made to move back and forth (horizontal scanning) and upward and downward (vertical scanning). In the actual tube, the beam starts at the top left-hand corner as in Fig. 11-19, and moves across the screen, left

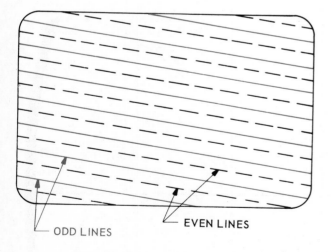

Fig. 11-19. The odd lines are traced first and then the even lines. There are 262 1/2 odd lines and 262 1/2 even lines in the total picture.

to right. At the extreme right position the beam is turned off and caused to FLY-BACK to the left position again. It is now moved downward slightly and made to trace another line in the same manner. This

beam traces out 15,750 horizontal lines per second. In American television 525 lines are used to form one complete picture and the picture is scanned 60 times per second. As the beam moves from top to bottom it first traces the lines 1, 3, 5 etc. or odd lines. When it reaches the bottom it rapidly returns to the top again and traces out the even numbered lines. This is referred to as INTERLACE SCANNING. It produces a better picture. One scan of the even or odd lines is a FIELD. Complete coverage of the screen by both even and odd lines is a FRAME. There will be 60 fields per second or 30 frames per second.

SYNCHRONIZATION OR SYNC

You will recall that the television camera at the studio also had an electron beam scanning a mosaic photo sensitive plate and converting the darkness of the elements into electrical signals. If the home TV is to reproduce this same picture, the scanning rate in the home TV must match the scanning rate of the TV camera. They must be in step like a squad of marching soldiers. To accomplish this feat, the TV transmitter sends a synchronization pulse to the home TV set for each line traced on the screen. This pulse keeps the horizontal and vertical sweep oscillator circuits exactly on frequency.

A signal required for two lines of some TV picture as transmitted over the air is drawn in Fig. 11-20. It is the composite VIDEO SIGNAL. It shows the varying voltages (for light or dark) applied to grid No. 1 of the picture tube as it traces one line on the screen. At the end of the line appears a BLANKING PULSE which turns off the beam during its retrace or return time. On top of the blanking pulse is the SYNC PULSE used to keep the home TV in step with the TV station.

Note that the greater the amplitude of the video information signal the blacker the spot on the screen appears. During the blanking pulse the screen is completely black. It happens too fast for the human eye to notice.

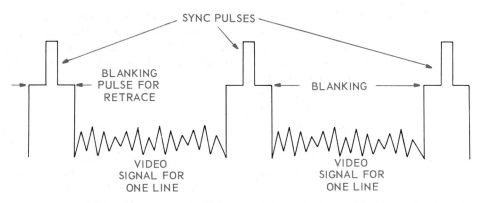

Fig. 11-20. The composite video signal showing video information, blanking pulse for retrace and sync pulse.

TV SOUND

The television station uses a typical FM transmitter for broadcasting the sound. This is an independent transmitter and it should always be in step with the picture. Once in a while something may happen like a dog barking with a soprano voice. Mistakes do happen.

EXPERIENCES WITH YOUR TV

There are two types of adjustments which are made on TV's. The first group are technical corrections and should be made by your serviceman. The second group are OWNER CONTROLS. These are the knobs on the front panel of the TV.

CHANNEL SELECTION. This is a ROTARY switch. Each channel is transmitted at a fixed frequency. The channel selector switches in the correct tank circuit (review tuning) to respond to the channel frequency.

FINE TUNING. This makes minor adjustments on the tuning circuit, so that your TV is exactly tuned to the incoming signals.

BRIGHTNESS CONTROL. This changes the grid voltage of the picture tube and makes a brighter or darker overall picture.

CONTRAST CONTROL. This changes the response of the amplifiers to make a greater difference between black and white.

HORIZONTAL HOLD. This control makes slight changes in the horizontal scanning oscillator frequency so that it will lock in step with the sync pulses.

VERTICAL HOLD. This adjusts the vertical scanning oscillator so that it will lock in step with the vertical sync signal.

FOCUS CONTROL. This controls voltages on picture tube to cause electron beam to converge to a narrow beam.

VOLUME CONTROL. Controls the level of sound. Usually the ON-OFF switch is part of this control.

COLOR TELEVISION

Color television produces a more realistic picture than monochrome (black and white) television. A color picture consists of the PRIMARY COLORS of RED, GREEN and BLUE in various combinations. All natural colors can be made by mixing these primary colors. In the television studio the camera separates the scene according to its color content by optical filters and sends the red, green and blue colors to separate camera tubes where they are scanned and converted into signals similar to the black and white camera.

These individual signals are not broadcasted as they are to your home, for a good reason. According to the FCC regulations, a black and white TV must be able to receive a color program and produce it as black and white. Therefore the individual color signals are combined to form a black and white signal and a composite color signal. These two signals are sent to the home receiver where they are separated to make the color picture. The black and white signal is made by combining the colors to indicate the variations in brightness of the picture. It is called the LUMINANCE SIGNAL. The colors in required proportions are mixed to form the CHROMINANCE SIGNAL. Both signals are used to modulate the transmitter. Multiplexing methods are used.

A color mixed with white can be washed out. On the other hand the red in our flag is very rich and vivid. Your color TV will have adjustments for HUE and SATURATION.

FORWARD STEPS IN UNDERSTANDING ELECTRONIC COMMUNICATIONS

1. Oscillation means swinging back and forth or to vibrate.
2. An oscillator produces an alternating wave which continuously varies through a cycle of values.
3. The peak voltage of a wave is its maximum instantaneous voltage.

PARTS LIST
L_1 Ferri Loopstick or Antenna Coil
C_1 365 pf Variable Capacitor
C_2 100 pf Ceramic Capacitor
D Crystal Diode 1N34

Fig. 11-21. A crystal detector.

COLOR TV RECEPTION

Your home TV receiver demodulates or detects the signals and decodes them. The three colors in proper proportions are fed to the picture tube. The color TV tube has three electron guns. One for each color. The screen of the color tube is phosphor coated with trios of red, blue and green phosphors. Each of the three color electron guns is directed to make its respective dot in the color trio glow. The total picture is a combination of all colors in varying degrees. When speaking of a color TV picture the term HUE is used. This is the color of the picture as the eye sees it. Grass has a green hue. Our flag has hues of red, blue and white. WHITE is a proper mixture of red, green and blue. SATURATION is an expression of how rich the color is.

4. The frequency of a wave is the number of cycles of variation per second. It is measured in HERTZ.
5. The period of a wave is the time in seconds for one cycle.
6. The effective value of an a-c wave is that value which produces the same heating power as a d-c current of that value. It is equal to .707 x Peak Value.
7. A tank circuit is used for tuning. Its frequency depends on the values of L and C.
8. Interrupted continuous waves can be used for communication by keying to the Morse Code.
9. When an audio tone is superimposed on a RF wave, it is called modulation.
10. Removing the audio tone from a RF signal is called demodulation or detection.
11. In AM communications the amplitude of

the RF signal is varied at an audio frequency rate.

12. In FM communications the frequency of the RF wave is varied at an audio frequency rate.
13. An AM detector is sensitive to amplitude variations in the modulated signal.
14. An FM detector is sensitive to frequency variations in the modulated signal.
15. In America, the TV picture is produced by using 525 lines of scanning.
16. The line frequency in TV is 15,750 lines per second.
17. In TV, 30 complete pictures of even and odd lines are sent each second.
18. The electron beam in a TV tube is moved back and forth by magnetic deflection coils.
19. TV transmitter and TV receiver are kept in step and exact sweep frequency by sync pulses sent from the transmitter.
20. The sound for TV is transmitted by FM.
21. Color television uses separate electron guns for red, green and blue.
22. Color phosphors on a TV screen are arranged in triangular trios of red, green and blue.
23. A color TV transmitter uses the luminance and chrominance signals for modulation.
24. Saturation means richness of color. Hue refers to the color of some object.

SAVE THE TRANSISTORS

Before making changes in a transistor circuit, always disconnect the power. This is also required if you remove a transistor from its socket. Voltage surges can destroy transistors.

TRANSISTORS CAN BE DESTROYED

When connected correctly, transistors are rugged little components. But if connected backwards with wrong voltage polarity, they can be destroyed. Check your connections BEFORE applying the power.

TEST YOUR KNOWLEDGE - UNIT 11

1. Build the tuning circuit and diode detector in Fig. 11-21. Connect earphones to the output. An antenna wire about ten feet long may be required. The antenna coil may be the commercial variety or about 40 turns of No. 20 wire wound on a tube. How many stations can you tune in? While listening to a station, remove C_2 from the circuit. Is the sound louder or weaker?
2. Build a ONE TRANSISTOR RADIO. The circuit and parts list is supplied in Fig. 11-22. What is the purpose of the transistor in this circuit? Which components

PARTS LIST
L_1 Ferri Loopstick
C_1 20-365 pf Variable Capacitor
C_2 .02 μf Disc Capacitor
C_3 .1 μf Paper Capacitor
D Diode 1N295
R_1 470 KΩ Resistor
Q_1 2N107 Transistor

Fig. 11-22. A transistor radio.

Fig. 11-23. The block diagram of detector and two transistor amplifier. Use shop power supply or transistor battery.

are used to tune in a station? What kind of a transistor is used? What is the purpose of C_2?

3. Build a RADIO DETECTOR AND AMPLIFIER. In this problem only the BLOCK DIAGRAMS will be supplied. Remember you studied about transistor amplifiers in an early chapter? Build the system as described in Fig. 11-23. Note that ground connections are assumed to be there. Why do you think they are omitted? Which component causes demodulation? What kind of transistors are used? Can circuit be tuned by moving core in L_1?

Unit 12

ELECTRONICS IN THE

MANUFACTURING INDUSTRIES

MAGNETISM AND MOTORS

The conversion of ELECTRICAL ENERGY TO MECHANICAL ENERGY is made possible by the ELECTRIC MOTOR. To understand the principle of the motor we need a basic understanding of the reaction between two magnetic fields. In Unit 7 on the Mystery of Magnetism we studied these laws of magnetism:

1. Unlike magnetic poles attract.
2. Like magnetic poles repel.

Later we discovered that a magnetic field existed around a current carrying conductor and the direction of this field depended on the direction of current flow.

CONDUCTOR MOTION

UP

FIELDS OPPOSE

FIELDS ADD

Fig. 12-1. *Above the conductor, the fields oppose and are weakened. The conductor moves in this direction.*

In Fig. 12-1 a current carrying conductor is shown in a magnetic field. Current is flowing "in" as indicated by the cross on the end of the wire. The circular magnetic field around the conductor illustrates this action. Above the conductor, the fixed field and the conductor field oppose each other. The conductor moves upward or toward the weakened field.

In Fig. 12-2 the opposite action takes place as current is flowing outward in the conductor as indicated by the dot. Now the

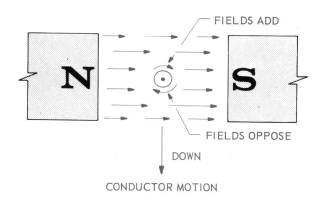

FIELDS ADD

FIELDS OPPOSE

DOWN

CONDUCTOR MOTION

Fig. 12-2. *The fields oppose below the conductor and are weakened. The conductor moves downward.*

field is weakened below the conductor and the conductor moves downward. This is called MOTOR ACTION. It is this principle which makes a motor rotate.

Fig. 12-3. Construction and theory of a simple d-c motor.

D-C MOTORS

A simple motor is sketched in Fig. 12-3. This has a permanent magnetic field. The conductor is in the shape of a coil, called the ARMATURE, and is mounted on bearings so it can rotate. Notice that the ends of the armature coil are connected to sections of a metal ring. This is the COMMUTATOR. It revolves with the armature. Outside power is brought into the armature through BRUSHES rubbing on the commutator sections.

In the first position follow the current flow from outside power through the brushes and commutator to the armature and out the other side. Current flows INWARD in armature coil marked C. Motor action causes an UPWARD movement. Current flows OUTWARD in armature coil marked D. Motor action causes a DOWNWARD movement. Therefore the coil turns.

After one-half revolution, the commutator section B comes in contact with brush A which reverses the armature current. Then side D moves upward and side C moves downward. The switching action of the commutator causes the armature to continue rotation.

To increase the turning power of the motor the armature may have many coils connected to commutator sections, and the field magnets may be electromagnets. A more practical motor is illustrated in Fig. 12-4.

THE MOTOR AS A GENERATOR

When you compare the simple d-c motor to the d-c generator in Unit 8, you will discover that they are quite similar. In fact, power applied to a generator will make it a motor. On the other hand, a motor is a generator. As the armature windings rotate through the magnetic field they generate a voltage which opposes the voltage from the source. This generator voltage is COUNTER ELECTROMOTIVE FORCE or CEMF.

We know that current in any circuit depends on the applied voltage. When a motor is not rotating there is no CEMF and the current is the result of armature resistance only and source voltage.

$$I_{armature} = \frac{E_{source}}{R_{armature}}$$

When the motor starts to rotate the generated CEMF opposes the source voltage. The net voltage becomes much less and the current decreases.

$$I_{armature} = \frac{E_{source} - E_{cemf}}{R_{armature}}$$

CONCLUSIONS: A motor draws a lot of current when starting. The current becomes less as the motor builds up speed. At home you may have noticed that your lights dim slightly when your refrigerator or other major appliance is turned on. Can you explain why?

A-C MOTORS

The majority of motors use alternating current. This power is available from your power company. Will the d-c motor operate on ac? With certain design changes it can become a pretty good motor. If the field and the armature change magnetic polarity at the same time, the motor action is the same. Motors used for both dc and ac are called UNIVERSAL MOTORS. These are used on small tools such as drill motors, grinders, and ventilating fans.

INDUCTION MOTORS

In the typical a-c induction motor the outside power is applied to the field or stator windings. By using inductance and capacitance with these windings a rotating magnetic field can be created. The armature

Fig. 12-4. Electromagnets are used for field coils. The armature has several coils connected to commutator sections.

Fig. 12-5. A commercial a-c induction motor. Note stator coils and construction of ROTOR.
(Delco Products, GMC)

or ROTOR is a laminated drum-like core with heavy single loop copper wires set in slots around its surface. Study Fig. 12-5 and notice this construction.

An induction motor can be considered a transformer. The stator windings will be the primary and the shorted loops of heavy copper in the rotor the secondary coils. The rotating magnetic field of the stator will induce a voltage in the rotor coil. The very low resistance of the single loop rotor coil will permit high currents to flow around the loops. These currents produce a magnetic field in

the rotor with a polarity which will cause the rotor to follow around and turn with the rotating stator field. Tremendous turning power or TORQUE can be developed in this manner.

TYPES OF A-C MOTORS

A-C induction motors are usually named according to the method used to create the revolving stator field. You will hear about SHADED POLE motors used in electric clocks and low-powered devices. The SPLIT PHASE motor uses starting and running wind-

ings of unequal inductances. The CAPACITOR·
START motor uses a capacitor to change
the phase of the ac to start the motor
rotating. The POLYPHASE INDUCTION MO-
TOR uses ac from a 3 phase source to
cause rotation.

WHAT IS PHASE?

A single source of a-c power may alternate at 60 cycles per second as illustrated in Fig. 12-6. A second current could
also be generated which would not rise and
fall in step with the first current. This
could be 2 phase. A more popular current
used for large motors in industry is the
3 phase. This current is generated by coils
in the generator 120 degrees apart. Each
set of coils will have its own external connection. Three or four wires are required
to conduct 3 phase current.

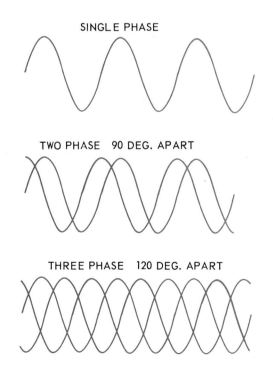

SINGLE PHASE

TWO PHASE 90 DEG. APART

THREE PHASE 120 DEG. APART

Fig. 12-6. The wave forms for single, two and three-phase
generators.

MEASURING RPM

Fig. 12-7 shows a rotating wheel of a
machine. Attached to the rim of the wheel

is a small permanent magnet. Each time the
wheel turns, the magnet passes a small
PICK UP coil. A voltage is induced in the
coil (Faraday's Discovery). If the pulses
occur in rapid succession as would happen

Fig. 12-7. A pulse is created for each revolution of wheel.
The meter is calibrated in rpm.

if the wheel revolved rapidly, an average
value of voltage could be induced. This can
be measured with a meter. If the speed of
the rotating wheel were increased, the average value of the induced voltage would increase. The meter, rather than reading voltage, could be marked in rpm (revolutions
per minute). A constant check on the speed
of the machine could be made. Such a circuit
can be used for measuring the speed of many
two cycle engines which have a magnet already on the flywheel.

A circuit similar to the rpm counter is
used in the newer automotive ignition systems. The coil and magnet are used instead
of breaker points in the distributor. The
pulse produced triggers a transistor or
solid state switch and the current in the
ignition coil. A high quality intensive spark
appears at the spark plug.

EXPERIENCE 1. Set up circuit pictured
in Fig. 12-8. A magnet is used for a core
of the coil. Move magnet in and out. Note
meter reading. Remove capacitor. What purpose did it serve. Reverse diode. Why is
diode used?

Fig. 12-8. *The pictorial sketch for* **EXPERIENCE 1.**

CONCLUSIONS: The increased movement produced a higher voltage. The meter could be read in "times per minute." The diode allows the current to flow in one direction only. The meter reads only in one direction. The capacitor seemed to average the voltage reading so that quick movements of the magnet did not cause the meter to move rapidly.

EXPERIENCE 2. Switching a transistor by a pulse. Build the circuit in Fig. 12-9. Move the magnet quickly in the coil. The lamp turns on. Does the transistor turn on when you move the magnet IN or OUT? Is a diode necessary in this circuit?

CONCLUSIONS: In order for the NPN transistor to conduct, the emitter must be negative and the base positive. This is forward bias for the NPN transistor. A voltage pulse of correct polarity will turn the transistor ON. A voltage pulse of opposite polarity would only turn it more OFF. The emitter-base junction is serving as a diode in this circuit.

TIMING CIRCUITS

There are many times in industry when timing an operation or process is required. Timing circuits are used extensively. Even at home you may use a timer when taking pictures with your camera. In the darkroom, photographs are enlarged. The interval of exposure time may be controlled by the electronic timer. Your mother may use a timer when boiling eggs for breakfast.

Most electronic timing circuits use the time constant of a RESISTANCE-CAPACITANCE circuit to set the time interval. The method is quite accurate for short periods of time when quality components are used.

Fig. 12-9. *The sketch for* **EXPERIENCE 2.** *The pulse from the magnet and the coil will turn on transistor. The lamp will light.*

EXPERIENCE 3. Build a VARIABLE TIMING CIRCUIT. The circuit is shown in Fig. 12-10. Pressing S_1 activates the circuit which energizes the relay and the indicator lamp turns ON. After an interval of time passes, adjusted by POTENTIOMETER R_2, the indicator lamp turns off.

CONCLUSIONS: In theory this circuit is not new; only the application is different. You will remember that the emitter-base of a transistor must be forward biased to cause current in the collector circuit. Therefore the BASE must be positive and the EMITTER negative for the NPN transistor to cause collector current.

The timer, in its inoperative condition, is held in an OFF state because C is charged

Fig. 12-10. The experimental TIMER circuit. The relay and lamp are used as an indicator.

PARTS LIST
R_1 – 10 KΩ Resistor
R_2 – 50 KΩ Potentiometer
C – 100 μf Capacitor
Q – 2N5449 Transistor
K – Relay – Sigma 11F2300 G

with the polarity indicated and the base of Q is negative. By pressing S_1 two circuit actions happen. First, C discharges since S_1 is a short circuit across C. Second, current flowing through R_1 makes the base end more positive and the transistor conducts. It will remain in conduction after S_1 is released until C regains its charge and makes the base negative.

EXPERIENCE 4. Use the SECOND hand of your watch. Measure the timing range of the circuit when R_2 is minimum and maximum resistance. Now experiment with the circuit. Substitute other values for R_2 and C. Can you explain the effect on the circuit by changing these components? You may wish to review RC time constants in Unit 9.

CONCLUSIONS: Although a relay and lamp are used in this circuit, the relay could be used to switch any type of electric motor or power device.

PHOTO CONTROL CIRCUITS

The use of light activated circuits further extends the science of electronics to control any type of machinery by varying the intensity of light. Photo devices are used to count products passing on an industrial conveyor. They are found in burglar alarm systems, safety and protective devices, camera light meters and thousands of other common and unusual circuits.

The conversion of light energy to electrical energy was discussed in Unit 4. This was the solar cell or sun cell. Also the component which changed its resistance was explained. It is a photoresistive cell.

EXPERIENCE 5. Build an ELECTRONIC COUNTER. The basic circuit is illustrated in Fig. 12-11. To "turn on" the transistor and activate the relay and lamp, the BASE of the transistor must be POSITIVE. When a light beam falls on the photoresistive cell, it causes it to conduct. This current causes the BASE end of R_1 to become positive and the circuit will operate.

CONCLUSIONS: The purpose of the potentiometer R_2 varies the total resistance of the base circuit and limits the current. Since the voltage at the base is the voltage drop across R_1 (E = IR) the R_2 will act as a SENSITIVITY CONTROL. If R_2 is set at a high resistance then a brighter light must be used to decrease the resistance of the photocell.

153

Fig. 12-11. The electronic counter circuit.

PARTS LIST
Photoresistive cell — Clairex C505
Transistor — TI 2N5449
Relay — Sigma 11F 2300 G SIL

If a numerical counter is attached in place of the indicator lamp, each time a light strikes the cell, it will be counted.

EXPERIENCE 6. Build a LIGHT OSCIL-LATOR. Change the circuit in Fig. 12-11 so that the indicating lamp is normally ON. In this connection, a light shining on the cell turns the indicator light OFF. Move the photocell near the indicator light. Now it will go crazy with flashing.

CONCLUSIONS: The indicator light ON turns the transistor ON which turns the light OFF which turns the transistor OFF and the light ON. This wild circuit cannot decide

just what it should do. Slight adjustments on R_2 may be required to make the circuit oscillate.

EXPERIENCE 7. Build a useful BURGLAR ALARM SYSTEM. The circuit combines the audio oscillator which you recently constructed in Unit 11 with a light control circuit to turn the audio tone ON or OFF depending on connections. The block diagram and relay connections are illustrated in Fig. 12-12. Team up with a friend. One of you may build the control; the other the oscillator.

CONCLUSIONS: It appears that a change

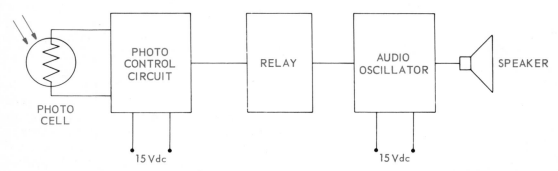

Fig. 12-12. The block diagram for the burglar alarm system. Separate power sources will be required if NPN transistors are used.

in light intensity may not only be measured but can be made to turn lights or machines ON or OFF as desired.

EXPERIENCE 8. MEASURE TURBIDITY. A turbid liquid is like muddy water. It is cloudy. Usually the turbidity of a liquid is the result of chemicals, dirt and solids suspended in the liquid. This is an important industrial test for water used in many processes. We will test turbidity by measuring the ability of a liquid to pass a beam of light. Build the simple circuit in Fig. 12-13. Shine a flashlight through the glass jar of water onto the photocell. Read the conduction current of the transistor. Record this meter reading as a reference. Now measure the turbidity of other liquids such as muddy water, milk and oil.

CONCLUSIONS: The meter is truly reading in milliamperes. However with a calibrated scale the meter could read in percentage of turbidity. Very sensitive circuits can detect extremely light changes in the clearness of the liquid.

TEMPERATURE MEASUREMENT

Temperatures have been measured for years in industry, science and at home. The use of electronic circuitry and amplifiers

Fig. 12-14. *A change of temperature produces a change of thermistor resistance and a change in current in the circuit.*

not only increases accuracy of measurement, but now permits the scientist to observe extremely small changes in temperature with the human eye. Changes can be measured and used to control other devices. A common example is in kitchen ovens which usually have automatic heat control. Basically a temperature measuring circuit will have a sensing device, an amplifier and a readout meter. In many industrial control systems the readout is compared to some standard and any difference is fed back to the heat source and corrects its operation.

EXPERIENCE 9. THERMISTORS. Build the simple thermistor circuit in Fig. 12-14. A thermistor is a solid state component which changes its resistance in proportion

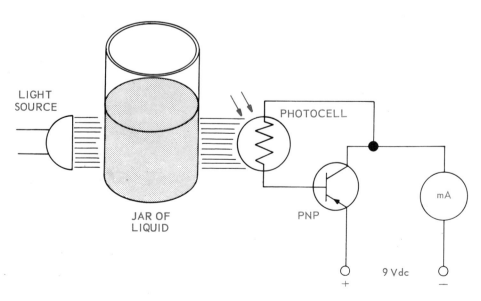

Fig. 12-13. *This circuit measures the turbidity or clearness of a liquid.*

to temperature. In this thermistor, a rise in temperature decreases its resistance and there is an increase in current. The meter can be calibrated to read in degrees rather than current.

ELECTRICAL ENERGY TO HEAT ENERGY

One of the most valuable effects of electricity is its ability to produce HEAT. When a current flows it must overcome the resistance of the wire conductor and any other resistive components. You might compare this to mechanical friction. If you use your brakes too much while driving down a steep hill in your car, the brake drums and linings may become very hot and even burn.

In many cases resistance is used in electronic circuits to limit current and the heat produced can be destructive. However, in this unit of study you will purposely use resistance to produce HEAT.

RESISTANCE AND WATTS

In Unit 4 you found that power used by an electrical device is equal to the current times the voltage and is measured in WATTS.

P (in watts) = I (in amperes) x E (in volts)

While performing your experiences in Unit 4 you also discovered the relationship between volts, amperes and ohms in a circuit. This was called OHM'S LAW and is stated mathematically as:

$$I \text{ (in amperes)} = \frac{E \text{ (in volts)}}{R \text{ (in ohms)}}$$

or

I (in amperes) x R (in ohms) = E (in volts)

Now look back at the power formula. It also uses E in volts. Since E is equal to I x R, we will just put I x R in the formula in place of E. The new formula will then read:

P (in watts) = I in amperes x (I x R)

or

$$\text{Power} = I^2 R$$

This is one of the more important relationships in all electricity. Power in WATTS EQUALS THE CURRENT SQUARED TIMES RESISTANCE.

A neat little poem is used in some of our Armed Services Schools to help you remember this formula:

"Twinkle, twinkle little star
Power equals I squared R"

NICHROME WIRE

A special wire (Nichrome) is manufactured which is an alloy of nickel and chromium. This has approximately 50 times more resistance than a copper wire of the same size. This and similar kinds of wire are used as HEATING ELEMENTS for stoves, toasters, heaters and many other places where heat is required. Nichrome wire will become red hot if enough current flows through it.

Fig. 12-15. *A heating element is made by winding Nichrome wire on a ceramic rod.*

AN ELECTRIC STOVE

Let's assume you wish to build a small electric stove which will use 550 watts of power when connected to your house electricity which is about 110 volts. The current necessary for this power is:

P = I x E   or   550 watts = I x 110 volts

I = 5 amperes

Next we need to compute the resistance which will permit this current to flow:

$$R = \frac{E}{I} \quad \text{or} \quad R \text{ (in ohms)} = \frac{110 \text{ volts}}{5 \text{ amperes}}$$

$$R = 22 \text{ ohms}$$

Nichrome wire has a resistance of 2 ohms per foot. To get 22 ohms, a piece 11 ft. long is necessary. This length of wire is wound in a loose coil around some kind of insulating, nonburning material.

EXPERIENCE 10. Secure from your shop supply a piece of No. 24 Nichrome Wire about two feet long. Wind it in a loose coil on some insulating ceramic material such as a glass rod as illustrated in Fig. 12-15. Using the ohmmeter, measure and record the resistance of the Nichrome coil.

Connect the resistance coil in the circuit displayed in Fig. 12-16.

Set voltage at one volt - observe current
Set voltage at two volts - observe current
Set voltage at five volts - observe current

Compute the power used at each voltage setting. CAUTION: THE RESISTANCE WIRE MAY GET RED HOT. DO NOT BURN YOUR-SELF OR START A FIRE.

CONCLUSIONS: As the voltage is increased the current also increases. The power increases as the square of the current. This means that if the current increases two times, the power increases two squared times or four times. Using a fixed voltage as found in your home, the power used by any light or appliance depends on the resistance of the light or appliance.

FORWARD STEPS IN UNDERSTANDING ELECTRICITY-ELECTRONICS

1. A motor converts electrical energy into mechanical energy.
2. The movement of a current carrying conductor in a magnetic field is MOTOR ACTION.
3. A d-c motor uses a commutator to change the direction of current in its armature.
4. An a-c induction motor has stator windings and a rotor which follows the rotating field of the stator.
5. A rotating a-c field is produced by out-of-phase currents in the stator coils.
6. A motor develops counter emf when it starts to rotate.
7. RPM can be measured by counting the pulses of voltage produced by a fixed magnet on the rotating wheel as it passes a pick-up coil.
8. Electronic timing circuits use the RC time constant of the circuit to set the time interval.

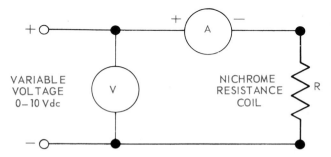

Fig. 12-16. The Nichrome heating element is connected in a circuit. Voltage and current can be observed.

9. Light intensity can be sensed by photo voltaic and photoresistive cells. The output of these cells can be used to control machinery.
10. A production line counter may use a photoelectric device.
11. Photoelectric devices can be used in counting and sorting machines, safety devices for machinery, burglar alarm systems and hundreds of other applications.
12. Photoelectric devices can be used to measure the turbidity of a liquid.
13. Thermistors change resistance in response to heat and can be used to measure temperature and control production processes.
14. Thermistors can be used to offset the effect of heat by changing the resistance of a circuit.
15. Resistance converts electrical energy to heat energy.
16. Power equals I x E.
17. Power equals I^2R.

LAMP
60W 110 VOLTS

D

20 μf 150V

10 K   1/2 W

4.7 K
1 W

RELAY

110 Vac

PARTS LIST
Relay — Sigma 11F 2300 G SIL
Photocell — Clairex CL 505
Diode — 200v PIV

Fig. 12-17. A suggested circuit for porch or garden lamp control.

18. Appliances and heating devices use resistance wire as a heating element.

TEST YOUR KNOWLEDGE - UNIT 12

1. The following appliances, operating on 110 volts a-c house current are rated by their power. What is the approximate resistance of each?

APPLIANCE	WATTAGE	RESISTANCE
Light Bulb	100w	_____
Toaster	550w	_____
Electric Heater	1000w	_____

2. Devise a circuit which will measure the ability of a wall surface to reflect light.

3. Devise a circuit which may be used to detect fire and sound an alarm.

4. Why does a motor get hot when used to turn a heavy load?

5. Devise a circuit which you might use to count students passing in the school hallway.

6. A photo enlarger requires a 5, 10, and 15 second exposure time. Devise a circuit which will electronically control the light.

7. Design the project and circuit for a front door or garden lamp which will automatically turn on at sunset and turn off at daybreak. See suggested circuit, Fig. 12-17.

APPENDICES

USING THE OSCILLOSCOPE

The oscilloscope is one of the basic instruments for electronic measurement. In many ways it works just like a TV set. However, instead of a complete picture, you will observe a solid green line across the face of the Cathode Ray Tube (CRT). The line will vary up and down vertically according to the signal being measured. Now refer to the illustration as we learn about the instrument controls.

INTENSITY. This control knob varies the brightness of the pattern on the face of the CRT. The ON-OFF switch is also a part of this control.

FOCUS. This makes the pattern a sharp, clearly defined line.

V CENTERING. This knob moves the green pattern up or down so that you may center it on the screen.

H CENTERING. This control moves the pattern to right or left for centering on the face of the CRT.

SWEEP. This is a switch which selects the frequency at which the spot on the face of the CRT moves from left to right.

SWEEP VERNIER. A control to vary the sweep frequency within the selected SWEEP range to the required exact frequency.

H GAIN. Adjusts the length of pattern across screen face.

V RANGE. A switch to set the oscilloscope to the range of peak-to-peak voltages to be measured.

A typical oscilloscope used in Radio and TV service as well as in laboratory experiences.

V CALIBRATE. To vary the gain of the vertical response within the selected V range. Used with,

V CAL. To calibrate the gain so that voltage can be read directly from scope pattern.

SYNC AND SYNC ADJUST. These are used to adjust a pulse from incoming signal to lock-in and produce a stationary pattern.

V INPUT and GND. These terminals will be used with test leads to apply signal to scope.

Let's assume you wish to observe and measure an a-c signal with a frequency of 400 Hz and an amplitude of 20 volts peak-to-peak. Use audio generator.

1. Turn scope on and adjust intensity and focus for a clear, sharp, green line across the face. (Use only brightness necessary for good pattern. A very bright line shortens the life of the CRT.)
2. Center the green line using V and H centering.
3. Adjust H GAIN until complete line is displayed.
4. Set V RANGE to 50 volt p-p scale.
5. Set Internal SYNC to +INT.
6. Connect the signal to be measured to V INPUT and GND.
7. Set SWEEP to 100-1000Hz range.
8. Adjust SWEEP VERNIER until two stationary sine wave patterns appear on the screen.

For the calibration of the scope so voltage peaks can be read directly, consult the MANUAL for your particular oscilloscope. Practice using the oscilloscope with other signals of different frequencies.

USING THE
AUDIO GENERATOR

You will use this type of instrument in many of your laboratory experiences. It is used by the technician for servicing and adjusting equipment operating at frequencies up to 200 KHz. The instrument is a SIGNAL SOURCE. It is easy to operate.

A typical Audio Frequency Signal Generator which produces both sine and square waves up to 200 KHz. (RCA)

LINE FREQUENCY OUTPUT. (Also ON-OFF switch.) See illustration. Adjusts the amplitude of a signal taken from the a-c line at 60 Hz. Output is taken from LINE FREQUENCY terminals.

OUTPUT TERMINALS. All signal outputs other than line frequency are taken from these terminals.

CONTROL AND DIAL. Use to set frequency of output signal. Use with,

FREQUENCY RANGE SWITCH. Dial reading times frequency range position will give output frequency. Example: Dial at 50. Range Switch at x10. Frequency output, 50 x 10 = 500 Hz.

ATTENUATOR. Determines amplitude of output wave in four steps. Generator can produce either SINE waves or SQUARE waves.

OUTPUT. A control of the amplitude of the output wave within any range of attenuator.

EXPERIENCE: Connect audio generator directly to the V INPUT of the oscilloscope. Set generator at 400 Hz. Adjust oscilloscope to display four sine waves about one inch high in amplitude. Change frequency of generator and/or oscilloscope and make adjustments for a stable pattern. Consult oscilloscope manual to learn about CALIBRATION so you may use the oscilloscope for direct measurement of generator output.

SIGNS AND SYMBOLS

DIODE, SEMICONDUCTOR

GROUND, EARTH

GROUND, CHASSIS

TRANSFORMER AIR CORE

TRANSFORMER IRON CORE

COILS, CHOKE AIR CORE

COIL, CHOKE IRON CORE

CAPACITOR FIXED VALUE

CAPACITOR ELECTROLYTIC

CAPACITOR VARIABLE

RESISTOR FIXED VALUE

RESISTOR TAPPED

RESISTOR POTENTIOMETER

VOLTAIC CELL

BATTERY

VOLTMETER

AMMETER

OHMMETER

THERMOCOUPLE

PHOTOVOLTAIC CELL

PHOTORESISTIVE CELL

FUSE

Appendices

AF GENERATOR	WIRES CROSS, CONNECTED	DIODE, VACUUM TUBE
LAMP	WIRES CROSS, NOT CONNECTED	JACK, PHONO
LAMP, NEON	RELAY COIL	RELAY, DOUBLE CONTACT
SWITCH SPST	RELAY	CRYSTAL
SWITCH SPDT	HEADPHONES	PHONO PICKUP
SWITCH DPST	SPEAKER	MICROPHONE
SWITCH DPDT	TRIODE	TRANSISTOR, PNP
PUSH BUTTON, Normally Open	TETRODE	TRANSISTOR, NPN
PUSH BUTTON, Normally Closed	PENTODE	SCR SILICON CONTROLLED RECTIFIER
ANTENNA		ZENER DIODE
CONNECTOR, PLUG IN		
CONNECTOR, RECEPTACLE		
TERMINAL, BINDING POST		

GREEK LETTER SYMBOLS

Ω - Omega   ohms
μ - mu   amplification factor, micro
$\propto$ - alpha   current gain, CB transistor
β - beta   current gain, CE transistor
λ - lambda   wavelength
π - Pi   3.1416
Δ - Delta   a change of

ABBREVIATIONS USED IN ELECTRONICS

R - RESISTANCE
L - INDUCTOR
C - CAPACITOR
X_L - INDUCTIVE REACTANCE
X_C - CAPACITIVE REACTANCE
Z - IMPEDANCE
E - VOLTAGE
V - VOLTAGE in Solid State Devices
Q or T - TRANSISTOR
V - VACUUM TUBE
CEMF - COUNTER ELECTROMOTIVE
        FORCE

A - GAIN
K - RELAY
RMS - ROOT MEAN SQUARE, Effective
       Value
T - TRANSFORMER
H - HENRY
F - FARAD
Hz - CYCLES PER SECOND

PREFIXES

milli (m) = 1/1000 of the unit 1×10^{-3}
micro (μ) = 1/1000000 of the unit 1×10^{-6}
pico   (p) = 1/1000000 of 1/000000 of unit
             1×10^{-12}
kilo   (K) = 1000 units 1×10^{3}
mega (M) = 1000000 units 1×10^{6}

EXAMPLES:

     800 KHz = 800000 Hz
      20   μf = .00002 f
    100   pf = .0000000001 f
      22   KΩ = 22000 Ω
       1   MΩ = 1000000 Ω
      10   mA = .01 A

CONSTRUCTION MATERIALS

There is an old saying, "Anything worth doing, is worth doing well." It is true that electronic circuits will often work if the component parts are placed helter-skelter and the interconnecting wires seem to resemble a bird's nest. However, this is not the mark of a good technician. Thought and order and neatness should show in any finished circuit or product. It is good advice to develop habits of skillful workmanship early in your electronic construction experiences.

There are many kinds, types and varieties of hardware used in the assembly of electronic equipment.

BECOMING FAMILIAR WITH SIZES

A bolt or machine screw may be specified as:

6-32 x 1/2 RH MACHINE SCREW

The first number 6 refers to its size in diameter. The second number 32 means 32 threads per inch. The third figure 1/2 designates its length in inches. RH means Round Head and Machine Screw tells you the type of screw.

SIZE	THREADS PER INCH	DIAMETER OF SCREW
2	56	.086 in.
4	36	.112 in.
4	40	.112 in.
6	32	.138 in.
8	32	.164 in.
10	32	.190 in.

TABLE 1. Screw Sizes.

Table 1 shows typical sizes of screws, threads per inch and diameter in inches. This information will help you select the proper screw and matching nut as well as the size drill to use in making screw holes.

These sizes are mostly used in electronic equipment. The screws used should be nickel plated. Your projects should not be put together with iron stove bolts.

TYPES OF HARDWARE

Soldering Lugs. (G. C. Electronics)

The lugs are used to make finished connections to parts and components. Wires may be crimped or soldered to the lug. These are sized by type, hole size and length.

Terminal Strips.

Terminal strips are made from one lug to eight lugs and more. These are used to mount small components and to connect several wires to one point. The lug is bolted to the chassis.

 Flat Metal Washer.

 Lock Washer.

 External Lock Washer.

 Shoulder Fibre Washer.

Washers.

Washers are used in many places. They are made in sizes to match machine screws. The shoulder washer is used with a flat fibre washer to insulate a binding post from a metal chassis.

Plugs and Jacks.

These plugs and jacks are used on the ends of test leads and connectors when a quick easy disconnect is required.

Binding posts are used when it is necessary to connect wires to a circuit panel. The

Binding Posts.

connecting wire is inserted in the hole and the top is screwed down. Some binding posts also have a banana plug jack in the top.

Battery and Alligator Clips.

Battery and alligator clips are used on the ends of wires and test leads for making quick connections to circuits.

 Round Head Screws.

Binding Head Screws.

Flat Head Screws.

Sheet Metal Screws.

Bolt and Screw Heads.

Screws are made in many sizes, lengths, head types and materials. Nylon screws will be found on some electronic equipment.

Spade Bolts.

Spade bolts are used to fasten components to panels and chassis.

Hexagon Nuts.

Nuts are ordered by size and threads per inch, also by width and thickness.

Appendices

Mounting Nuts.

Mounting nuts are used to mount controls, switches and other parts to the front panel of the equipment.

Fahnestock Clips.

Fahnestock clips are used to make temporary connections of wires to a circuit or board.

Grommets.

Grommets are used to protect and insulate wires passing through a hole in a metal chassis.

Only a small sample of electronic hardware has been illustrated. These common ones will be needed in making simple projects.

ACKNOWLEDGMENTS

To produce this text on exploration and discovery in electronics required the efforts of many people from education and industry.

Thanks are given to the author's wife, Virtue B. Gerrish, for editing and typing the manuscript and to Perry Page, an industrial arts major at Humboldt State College, for preparing illustrations, building and checking experiences and activities.

The author wishes to express his sincere thanks and appreciation to those who have contributed so generously to complete the manuscript, illustrations and activities, as listed below.

American Radio Relay League
Ampex Corp.
Bell Laboratories
Bernback Co.
Boeing Aircraft Corp.
Buck Engineering Co.
Bulova Watch Co.
Bussman Mfg. Co.
Centralab – Globe Union, Inc.
Central Scientific Co.
CBC Industries
Clevite Corp., Burgess Battery Division
Delco Products, General Motors Corp.
Delco-Remy, General Motors Corp.
Eastman Kodak Co.
EDUCOMP
General Electric Co.
Graymark Enterprises, Inc.
Gulton Industries, Inc.
Hewlett Packard Co.
International Rectifier Corp.
IRC, Inc.
J. W. Miller Co.
KHSL, Channel 12, Chico, California

KIRSH Co.
Lepel Laboratories
Lindberg Engineering Co.
LTV – University Sound
Minneapolis Honeywell Regulator Co.
Motorola Semiconductor Products
NASA
Ohmite Mfg. Co.
Pacific Gas and Electric Co.
Pacific Telephone Co.
Philco–Ford Corp.
P. R. Mallory Co.
Radar Devices Mfg. Corp.
Radio Corp. of America
Rathcon, Inc.
Sprague Electric Co.
Square D. Co.
Triad Transformer Co.
Triplett Electrical Instrument Co.
Union Carbide Consumer Products
Union Switch and Signal Co.
Westinghouse Electric Co.
Mr. Bob Fikes, Electronics Instructor in
    Alta Loma High School, California

LANGUAGE OF INDUSTRY

FOR ELECTRICITY-ELECTRONICS

AC: Alternating current.

AC GENERATOR: Generator using slip rings and brushes to connect armature to external circuit. Output is alternating current.

AGC: Abbreviation for AUTOMATIC GAIN CONTROL. Circuit employed to vary gain of amplifier in proportion to input signal strength so output remains at constant level.

AIR-CORE INDUCTOR: Inductor wound on insulated form without metallic core. Self-supporting coil without core.

ALIGNMENT: The adjustment of tuned circuits in amplifier and/or oscillator circuits so that they will produce a specified response at a given frequency.

ALNICO: Special alloy used to make small permanent magnets.

ALPHA: Greek letter α, represents current gain of a transistor. It is equal to change in collector current caused by change in emitter current for constant collector voltage.

ALTERNATING CURRENT (AC): Current of electrons that moves first in one direction and then in the other.

ALTERNATOR: An AC generator.

AMMETER: Meter used to measure current.

AMPERE: This unit measures electricity "on the move" or flowing in a circuit. Moving electricity is called current. It is this movement of electrical energy which does the work. It produces heat for your electric stove and light for your home. It produces music from your Hi-Fi Stereo and pictures and sound from television. Scientists have agreed that if a certain quantity of electricity passes a given point in a circuit in one second, it will be called one ampere. Smaller units than an ampere are:

MILLIAMPERE - one thousandth of an ampere or .001 ampere.

MICROAMPERE - one millionth of an ampere or .000001 ampere.

AMPERE-HOUR: Capacity rating measurement of batteries. A 100 ampere-hour battery will produce, theoretically, 100 amperes for one hour.

AMPERE TURN (IN): Unit of measurement of magnetomotive force. Represents product of amperes times number of turns in coil of electromagnet. F = 1.257 IN.

AMPLIFICATION: Ability to control a relatively large force by a small force. In a vacuum tube, relatively small variation in grid input signal is accompanied by relatively large variation in output signal.

AMPLIFICATION FACTOR: Expressed as μ (mu). Characteristic of vacuum tube to amplify a voltage. Mu is equal to change in plate voltage as result of change in grid voltage while plate current is constant.

AMPLIFIERS: Power. Electron tube used to increase power output. Sometimes called a current amplifier.

VOLTAGE. An electron tube used to amplify a voltage.

AF AMPLIFIER. Used to amplify audio frequencies.

IF AMPLIFIER. Used to amplify intermediate frequencies.

RF AMPLIFIER. Used to amplify audio frequencies.

AMPLITUDE: Extreme range of varying quantity. Size, height of.

AMPLITUDE MODULATION (AM): Modulating a transmitter by varying strength of r-f carrier at audio rate.

AND CIRCUIT: A circuit with two or more inputs and all inputs must be present to produce an output signal.

ANODE: Positive terminal, such as plate in electron tube.

ANTENNA: Device for radiating or receiving radio waves.

ARMATURE: Revolving part in generator or motor. Vibrating or moving part of relay or buzzer.

ATOM: Smallest particle that makes up a type of material called an element.

ATOMIC NUMBER: Number of protons in nucleus of a given atom.

ATOMIC WEIGHT: Mass of nucleus of atom in reference to oxygen which has a weight of 16.

ATTENUATOR: Decrease in amplitude or intensity.

AUTOMATION: An industrial technique of employing automatic self-controlled machinery to replace human labor and control.

AVC: Automatic volume control.

AVERAGE VALUE: Value of alternating current or voltage of sine wave form that is found by dividing area under one alternation by distance along X axis between 0 and 180 deg. $E_{avg} = .637 E_{max}$.

AWG: AMERICAN WIRE GAUGE - used in sizing wire by numbers.

BAND: Group of adjacent frequencies in frequency spectrum.

BAND SWITCHING: Receiver employing switch to change frequency range of reception.

BANDWIDTH: Band of frequencies allowed for transmitting modulated signal.

BASE: The thin section between the emitter and collector of a transistor.

BATTERY: Several voltaic cells connected in series or parallel. Usually contained in one case.

BETA: Greek letter β, represents current gain of common-emitter connected transistor. It is equal to ratio of change in collector current to change in base current, while collector voltage is constant.

BIAS, FORWARD: Connection of potential to produce current across PN junction. Source potential connected so it opposes potential hill and reduces it.

BIAS, REVERSE: Connection of potential so little or no current will flow across PN junction. Source potential has same polarity as potential hill and adds to it.

BINARY: Number system having base of 2, using only the symbols 0 and 1.

BLEEDER: Resistor connected across power supply to discharge filter capacitors.

BRIDGE CIRCUIT: Circuit with series-parallel groups of components that are connected by common bridge. Bridge is frequently a meter in measuring devices.

BRIDGE RECTIFIER: Full-wave rectifier circuit employing four rectifiers in bridge configuration.

BRIDGE, WHEATSTONE: A bridge circuit for determining the value of an unknown component by comparison to one of known value.

BRUSH: Sliding contact, usually carbon, between commutator and external circuit in d-c generator.

B-SUPPLY: Voltage supplied for plate circuits of electron tubes.

BYPASS CAPACITOR: Fixed capacitor which bypasses unwanted ac to ground.

CABLE: May be stranded conductor or group of single conductors insulated from each other.

CAPACITANCE: Inherent property of electric circuit that opposes change in voltage. Property of circuit whereby energy may be stored in electrostatic field.

CAPACITIVE REACTANCE (X_C): The opposition to the flow of an alternating current as the result of counter voltages stored in the capacitor when a varying voltage is applied. It is measured in OHMS and is inversely proportional to the frequency of applied voltage and the value of the capacitor in farads.

CAPACITOR: Device which possesses capacitance. Simple capacitor consists of two metal plates separated by insulator.

CAPACITY: Ability of battery to produce current over given length of time. Capacity is measured in ampere-hours.

CARRIER: Usually radio frequency continuous wave to which modulation is applied. Frequency of transmitting station.

CARRIER (in a semiconductor): Conducting hole or electron.

CASCADE: Arrangement of amplifiers where output of one stage becomes input of next, throughout series of stages.

CATHODE: Emitter in electron tube.

CATHODE RAY TUBE: Vacuum tube in which electrons emitted from cathode are shaped into narrow beam and accelerated to high velocity before striking phosphor-coated viewing screen.

CENTER FREQUENCY: Frequency of transmitted carrier wave in FM when no modulation is applied.

CENTER TAP: Connection made to center of coil.

CHOKE COIL: A high inductance coil used to prevent the passage of pulsating currents, but allows dc to pass.

CHOKE, RF: A choke coil with a high impedance at radio frequencies.

CIRCUIT BREAKER: Safety device which automatically opens circuit if overloaded.

CIRCUIT, ETCHED: The method of circuit board production in which the actual conduction paths on a copper-clad insulation board are coated with an acid resist. The board is then placed in an acid bath and unprotected parts of the copper-clad are eaten away, leaving the circuit conductors. Components are mounted and soldered between the conductors to form the completed circuit.

CIRCUIT, PRINTED: The method of printing circuit conductors on an insulated base. Component parts may also be printed or actual components soldered in place.

CITIZENS' BAND: A band of frequencies allotted to two-way radio communications by private citizens. Operators are not required to pass technical examinations.

COLLECTOR: In a transistor, the semiconductor section which collects the majority carriers. Similar to the plate in a vacuum tube.

COMMON BASE: Transistor circuit, in which base is common to input and output circuits.

COMMON COLLECTOR: Transistor circuit in which collector is common to input and output circuits.

COMMON EMITTER: Transistor circuit in which emitter is common to input and output circuits.

COMMUTATION: The process of changing the alternating current in a generator armature into direct current in the external circuit by a mechanical switch consisting of commutator bars and brushes.

COMMUTATOR: Group of bars providing connections between armature coils and brushes. Mechanical switch to maintain current in one direction in external circuit.

COMPUTER: ANALOG. A computer which substitutes for any given physical quantity, a mechanical, electrical thermodynamic equivalent quantity that follows in direct proportion, the same laws of behavior as the original quantity. In general, the analog computer gives a continuous solution of the problem. Ex: Slide rule, automobile speedometer.

DIGITAL. A computer which makes a one-to-one comparison or individual count to calculate. It solves problems in discrete steps, forming a discontinuous solution.

CONDUCTANCE (symbol G): Ability of circuit to conduct current.

It is equal to amperes per volt and is measured in mhos. $G = \dfrac{1}{R}$

CONDUCTIVITY: The ability of a material to conduct an electric current. It is the reciprocal of resistivity.

CONDUCTIVITY, N TYPE: Conduction by electrons in N-type crystal.

CONDUCTIVITY, P TYPE: Conduction by holes in a P-type crystal.

CONDUCTOR: Material which permits free motion of large number of electrons.

CONTINUOUS WAVE (CW): Uninterrupted sinusoidal rf wave radiated into space, with all wave peaks equal in amplitude and evenly spaced along time axis.

CONTROL GRID: Grid in vacuum tube closest to cathode. Grid to which input signal is fed to tube.

COPPER LOSSES: Heat losses in motors, generators and transformers as result of resistance of wire.

CORE IRON: Magnetic materials usually in sheet form, used to form laminated cores for electromagnets and transformers.

COULOMB: A specified quantity of electrons. The amount of electricity moving in a circuit in one second by a current of one ampere. A Coulomb, designated by the letter Q, equals a charge of 6.25×10^{18} electrons.

COUNTER EMF (cemf): Voltage induced in conductor moving through magnetic field which opposes source voltage.

COVALENT BOND: Atoms joined together, sharing each other's electrons to form stable molecule.

CROSSOVER NETWORK: The network designed to divide audio frequencies into bands for distribution to speakers.

CRYOGENICS: The use of electronic circuits designed to take advantage of increased efficiency at extremely low temperatures.

CRYSTAL DIODE: Diode formed by small semiconductor crystal and cat whisker.

CRYSTAL LATTICE: Structure of material when outer electrons are joined in covalent bond.

CURRENT: Transfer of electrical energy in conductor by means of electrons moving constantly and changing positions in vibrating manner.

CW: Abbreviation for Continuous Wave.

CYBERNETICS: The study of complex electronic computer systems and their relationship to the human brain.

CYCLE: A chain of events occurring in sequence. A voltage might start at zero and rise to a peak value in a positive direction, return to zero and then rise to peak value in a negative direction and then return to zero. One cycle is complete. Usually cycles continue to repeat this sequence of events.

D'ARSONVAL METER: Stationary-magnet moving coil meter.

DC: Direct current.

DC GENERATOR: Generator with connections to armature through a commutator. Output is direct current.

DCWV: Abbreviation for DIRECT CURRENT WORKING VOLTAGE. It is a specification of a capacitor.

DECAY: Term used to express gradual decrease in values of current and voltage.

DECAY TIME: The time required for a capacitor to discharge to a specified percentage of its original charge.

DECIBEL: One-tenth of a Bel. A unit used to express the relative increase or decrease in power. Unit used to express gain or loss in a circuit.

DEFLECTION: Deviation from zero of needle in meter. Movement or bending of an electron beam.

DEMODULATION: Process of removing modulating signal intelligence from carrier wave in radio receiver.

DETECTION: See DEMODULATION.

DETECTOR: CRYSTAL. A type of detection which uses the rectification characteristics of a crystal substance such as galena, silicon, germanium, iron pyrite.

DIODE: A detector using a diode tube or semiconductor as the rectifier of the rf signal.

DIAPHRAGM: Thin disc, used in an earphone for producing sound.

DIELECTRIC: Insulating material between plates of capacitor.

DIODE: A two element vacuum tube or semiconductor. A diode conducts in the forward direction but offers high resistance to currents in the opposite direction.

DIODE DETECTOR: Detector circuit utilizing unilateral conduction characteristics of diode.

DIRECT CURRENT (DC): Flow of electrons in one direction.

DOPING: Adding impurities to semiconductor material.

DRY CELL: Nonliquid cell, which is composed of zinc case, carbon positive electrode and paste of ground carbon, manganese dioxide and ammonium chloride as electrolyte.

DYNAMIC SPEAKER: Speaker which produces sound as result of reaction between fixed magnetic field and fluctuating field of voice coil.

EDISON EFFECT: Effect, first noticed by Thomas Edison, that emitted electrons were attracted to positive plate in vacuum tube.

EDDY CURRENTS: Induced current flowing in rotating core.

EDDY CURRENT LOSS: Heat loss resulting from eddy currents flowing through resistance of core.

EFFECTIVE VALUE: That value of alternating current of sine wave form that has equivalent heating effect of a direct current. $(.707 \times E_{peak})$

EFFICIENCY: Ratio between output power and input power.

ELECTRODE: Elements in a cell.

ELECTRODYNAMIC SPEAKER: Dynamic speaker that uses electromagnetic fixed field.

ELECTROLYTE: Acid solution in a cell.

ELECTROLYTIC CAPACITOR: Capacitor with positive plate of aluminum. Dry paste or liquid forms negative plate. Dielectric is thin coat of oxide on aluminum plate.

ELECTROMAGNET: Coil wound on soft iron core. When current runs through coil, core becomes magnetized.

ELECTROMOTIVE FORCE (emf): Force that causes free electrons to move in conductor. Unit of measurement is the volt.

ELECTRON: Negatively charged particle.

ELECTRON TUBE: Highly evacuated metal or glass shell which encloses several elements.

ELECTROSTATIC FIELD: Space around charged body in which its influence is felt.

ELEMENT: One of the distinct kinds of substances which either singly or in combination with other elements, makes up all matter in the universe.

EMISSION: Escape of electrons from a surface.

EMISSION, Types of: PHOTOELECTRIC. Emission of electrons as result of light striking surface of certain materials.

SECONDARY. Emission caused by impact of other electrons striking surface.

THERMIONIC: Process where heat produces energy for release of electrons from surface of emitter.

EMITTER: Element in a vacuum tube from which electrons are emitted. The CATHODE.

EMITTER: MAJORITY. In a transistor, the semiconductor section, either P or N type, which emits majority carries into the interelectrode region.

MINORITY. In a transistor, the semiconductor section, either P or N type, which emits minority carriers into the inter-electrode region.

ENERGY: That which is capable of producing work.

EXCITATION: To apply a signal to an amplifier circuit; to apply energy to an antenna system; to apply current to energize the field windings of a generator.

FARAD: This unit describes the capacity or size of a capacitor. It is the ability of a component to store an electric charge. It is agreed that a capacitor which stores a specified quantity of electrons, called a COULOMB, when a one volt electric force is applied to it, will have a value of one farad. The farad is a large unit and smaller units used in electronics are:

MICROFARAD - one millionth of a farad or .000001 farad.

PICOFARAD - one millionth of one millionth of a farad or .000000000001 farad. (Used to be called a micromicrofarad.)

FEEDBACK: Transferring voltage from output of circuit back to its input.

FIELD MAGNETS: Electromagnets which make field of motor or generator.

FILTER: Circuit used to attenuate specific band or bands of frequencies.

FLUX: Represented by Greek letter ϕ; total number of lines of magnetic force.

FLUX DENSITY (symbol B): Number of lines of flux per cross-sectional area of magnetic circuit.

FREQUENCY: The rate of number of times per second that a cycle occurs. Frequency is measured in "cycles per second" or HERTZ. These terms will also be used:

KILOHERTZ (KHz) = 1,000 cycles per second.

MEGAHERTZ (MHz) = 1,000,000 cycles per second.

FREQUENCY BANDS: VLF - Very low frequencies, 10-30 KHz.
  LF - Low frequencies, 30-300 KHz.
  MF - Medium frequencies, 300-3,000 KHz.
  HF - High frequencies, 3-30 MHz.
  VHF - Very high frequencies, 30-300 MHz.
  UHF - Ultra high frequencies, 300-3,000 MHz.
  SHF - Super high frequencies, 3,000-30,000 MHz.
  EHF - Extremely high frequencies, 30,000-300,000 MHz.

FREQUENCY DEPARTURE: Instantaneous change from center frequency in FM as result of modulation.

FREQUENCY DEVIATION: Maximum departure from center frequency at the peak of the modulating signal.

FREQUENCY MODULATION (FM): Modulating transmitter by varying frequency of r-f carrier wave at an audio rate.

FREQUENCY RESPONSE: Rating of device indicating its ability to operate over specified range of frequencies.

FREQUENCY SWING: The total frequency swing from maximum to minimum. It is equal to twice the deviation.

FULL-WAVE RECTIFIER: Rectifier circuit which produces a dc pulse output for each half cycle of applied alternating current.

FUNDAMENTAL: A sine wave that has the same frequency as complex periodic wave. Component tone of lowest pitch in complex tone. Reciprocal of period of wave.

FUSE: Safety protective device which opens an electric circuit if overloaded. Current above rating of fuse will melt fusible link and open circuit.

GAIN: Ratio of output a-c voltage to input a-c voltage.

GALVANOMETER: Meter which indicates very small amounts of current and voltage.

GATE: A circuit which permits an output only when a predetermined set of input conditions are met.

GAUSS: Measurement of flux density in lines per square centimeter.

GENERATOR: Rotating electric machine which provides a source of electrical energy. A generator converts mechanical energy to electric energy.

GENERATORS, Types of: COMPOUND. Uses both series and shunt winding.

INDEPENDENTLY EXCITED. Field windings are excited by separate d-c source.

SERIES. Field windings are connected in series with armature and load.

SHUNT. Field windings are connected across armature in shunt with load.

GERMANIUM: A rare grayish-white metallic chemical element. Symbol Ge; Atomic Wgt. 72.60; Atomic Number 32.

GILBERT: Unit of measurement of magnetomotive force. Represents force required to establish one maxwell in circuit with one Rel of reluctance.

GRID: Grid of fine wire placed between cathode and plate of an electron tube.

GRID BIAS: Voltage between the grid and cathode, usually negative.

GRID DIP METER: A test instrument for measuring resonant frequencies, detecting harmonics and checking relative field strength of signals.

HALF-WAVE RECTIFIER: Rectifier which permits one-half of an alternating current cycle to pass and rejects reverse current of remaining half-cycle. Its output is pulsating dc.

HARMONIC FREQUENCY: Frequency which is multiple of fundamental frequency. Example: If fundamental frequency is 1000 KHz, then second harmonic is 2 x 1000 KHz or 2000 KHz; third harmonic is 3 x 1000 or 3000 KHz, and so on.

HEATER: Resistance heating element used to heat cathode in vacuum tube.

HEAT SINK: Mass of metal used to carry heat away from component.

HENRY: This unit describes the value of the inductance of a coil. If the current through a coil is made to vary, a voltage is produced by the coil. This is called INDUCTANCE. If the current is changing at the rate of one ampere per second and a voltage of one volt is induced, the coil is said to have an inductance of one Henry. Smaller values are also used in electronics:

MILLIHENRY - one thousandth of a henry or .001 henry.

MICROHENRY - one millionth of a henry or .000001 henry.

HERTZ: The new unit of measurement for frequency, named in honor of Heinrich Hertz who discovered radio waves. One HERTZ equals "one cycle per second."

HETERODYNE: The process of combining two signals of different frequencies to obtain the difference frequency.

HOLE: Positive charge. A space left by removed electron.

HOLE INJECTION: Creation of holes in semiconductor material by removal of electrons by strong electric field around point contact.

HORSEPOWER: 33,000 ft. lbs. of work per minute or 550 ft. lbs. of work per second equals one horsepower. Also, 746 watts = 1 HP.

HUM: Form of distortion introduced in an amplifier as result of coupling to stray electromagnetic and electrostatic fields or insufficient filtering.

HYDROMETER: Bulb-type instrument used to measure specific gravity of a liquid.

HYSTERESIS LOSS: Energy loss in substance as molecules or domains move through cycle of magnetization. Loss due to molecular friction.

IMPEDANCE MATCHING: Sometimes called Z match; the matching of two different impedances to obtain maximum transfer of power.

IMPEDANCE (Z): The vector sum of the resistance, inductive reactance and the capacitive reactance in a circuit. R, X_C and X_L cannot be added directly since both the opposition in ohms of X_L and X_C are not in phase with R.

IMPURITY: Atoms within a crystalline solid which are foreign to the crystal.

INDUCED CURRENT: Current that flows as result of an induced emf.

INDUCED EMF: Voltage induced in conductor as it moves through magnetic field.

INDUCTANCE: Inherent property of electric circuit that opposes a change in current. Property of circuit whereby energy may be stored in magnetic field.

INDUCTANCE, MUTUAL: See MUTUAL INDUCTANCE.

INDUCTION MOTOR: AC motor operating on principle of rotating magnetic field produced by out-of-phase currents. Rotor has no electrical connections, but receives energy by transformer action from field windings. Motor torque is developed by interaction of rotor current and rotating field.

INDUCTIVE REACTANCE (X_L): The opposition to the flow of an alternating current as the result of counter voltages induced in the coil by a varying current. It is measured in OHMS and is directly proportional to the frequency of the applied voltage and the value of the coil in henrys.

INDUCTOR: A coil; a component with the properties of inductance.

IN PHASE: Two waves of the same frequency are in phase when they pass through their maximum and minimum values at the same instant with the same polarity.

INPUT IMPEDANCE: The impedance of the input terminals of a circuit or device, with the input generator disconnected.

INSTANTANEOUS VALUE: Any value between zero and maximum depending upon instant selected.

INSULATORS: Substances containing very few free electrons and requiring large amounts of energy to break electrons loose from influence of nucleus.

INTERNAL RESISTANCE: Refers to internal resistance of source of voltage or emf. A battery or generator has internal resistance which may be represented as a resistor in series with source.

INTERRUPTED CONTINUOUS WAVE (ICW): Continuous wave radiated by keying transmitter into long and short pulses of energy (dashes and dots) conforming to code such as Morse Code.

INTERSTAGE: Existing between stages, such as in interstage transformer between two stages of amplifiers.

ION: An atom which has lost or gained some electrons. It may be positive or negative depending on the net charge.

IONIZATION: An atom is said to be ionized when it has lost or gained one or more electrons.

IONOSPHERE: Atmospheric layer from 40 to 350 miles above the earth, containing a high number of positive and negative ions.

JUNCTION DIODE: PN junction having unidirectional current characteristics.

JUNCTION TRANSISTOR: Transistor consisting of thin layer of N or P type crystal between P or N type crystals. Designated as NPN or PNP.

KEY: Manually operated switch used to interrupt r-f radiation of transmitter.

KEYING: Process of causing CW transmitter to radiate an r-f signal when key contacts are closed.

KILO: Prefix meaning one thousand times.

KILOWATT-HOUR (KWH): Means 1,000 watts per hour. Common unit of measurement of electrical energy for home and industrial use. Power is priced by the KWH.

LAMBDA: Greek letter λ. Symbol for wavelength.

LAMINATIONS: Thin sheets of steel used in cores of transformers, motors and generators.

LAWS OF MAGNETISM: Like poles repel; unlike poles attract.

LEAD ACID CELL: Secondary cell which uses lead peroxide and sponge lead for plates, and sulfuric acid and water for electrolyte.

LEFT-HAND RULE: A method, using your left hand, to determine polarity of an electromagnetic field or direction of electron flow.

LENZ'S LAW: Induced emf in any circuit is always in such a direction as to oppose effect that produces it.

LINEAR: In a straight line; a mathematical relationship in which quantities vary in direct proportion.

LINEAR AMPLIFIER: An amplifier whose output is in exact proportion to its input.

LINEAR DEVICE: Electronic device or component whose current-voltage relation is a straight line.

LINES OF FORCE: Graphic representation of electrostatic and magnetic fields showing direction and intensity.

LOAD: Resistance connected across circuit which determines current flow and energy used.

LOADING A CIRCUIT: Effect of connecting voltmeter across circuit. Meter will draw current and effective resistance of circuit is lowered.

LOCAL OSCILLATOR: Oscillator in superheterodyne receiver, output of which is mixed with incoming signal to produce intermediate frequency.

LODESTONE: Natural magnet, so called a "leading stone" or lodestone because early navigators used it to determine directions.

LOGICAL FUNCTIONS: An expression referring to a definite state of condition.

AND FUNCTION. An output is obtained only with a combined group of input signals.

OR FUNCTION. An output is obtained with any one of a group of input signals.

NOT FUNCTION. An output is obtained only when there is no input signal.

MEMORY FUNCTION: An output is continually obtained unless an input signal is applied.

LOUDSPEAKER: Device to convert electrical energy into sound energy.

L PAD: A combination of two variable resistors, one in series and the other across the load, which is used to vary the output of an audio system and match impedances.

MAGNET: Substance that has the property of magnetism.

MAGNETIC CIRCUIT: Complete path through which magnetic lines of force may be established under influence of magnetizing force.

MAGNETIC FIELD: Imaginary lines along which magnetic force acts. These lines emanate from N pole and enter S pole, forming closed loops.

MAGNETIC FLUX (symbol Φ phi): Entire quantity of magnetic lines surrounding a magnet.

MAGNETIC LINES OF FORCE: Magnetic line along which compass needle aligns itself.

MAGNETIC MATERIALS: Materials such as iron, steel, nickel and cobalt which are attracted to magnet.

MAGNETIC PICKUP: Phono cartridge which produces an electrical output from armature in magnetic field. Armature is mechanically connected to reproducing stylus.

MAGNETIZING CURRENT: Current used in transformer to produce transformer core flux.

MAGNETOMOTIVE FORCE (F) (mmf): Force that produces flux in magnetic circuit.

MAGNET POLES: Points of maximum attraction on a magnet; designated as north and south poles.

MAJOR CARRIER: Conduction through semiconductor as a result of majority of electrons or holes.

MATTER: Physical substance of common experience. Everything about us is made up of matter.

MAXIMUM VALUE: Peak value of sine wave either in positive or negative direction.

MEGA: Prefix meaning one million times.

MHO: Unit of measurement of conductance.

MICA CAPACITOR: Capacitor made of metal foil plates separated by sheets of mica.

MICRO: Prefix meaning one millionth of.

MICROMICRO: Prefix meaning one millionth of one millionth of.

MICROPHONE: Energy converter that changes sound energy into corresponding electrical energy.

MIL: One thousandth of an inch. (.001 inch)

MILLI: Prefix meaning one thousandth of.

MILLIAMMETER: Meter which measures in milliampere range of currents.

MINOR CARRIER: Conduction through semiconductor opposite to major carrier. Ex: If electron is major carrier, then hole is minor carrier.

MINUS (symbol -): Negative terminal of junction of circuit.

MODULATION: Process by which amplitude or frequency of sine wave voltage is made to vary according to variations of another voltage or current called modulation signal.

MODULATION PRODUCT: Sideband frequencies resulting from modulation of a radio wave.

MOLECULE: Smallest division of matter. If further subdivision is made, matter will lose its identity.

MOTOR: Device which converts electrical energy into mechanical energy.

MOTORS, Types of dc: COMPOUND. Uses both series and parallel field coils.

SERIES. Field coils are connected in series with armature circuit.

SHUNT. Field coils are connected in parallel with armature circuit.

MU: Greek letter (μ) used to represent the amplification factor of a vacuum tube; magnetic permeability; the prefix meaning one millionth of.

MULTIMETER: A combination volt, ampere and ohmmeter.

MULTIPLIER: Resistance connected in series with meter movement to increase its voltage range.

MULTIVIBRATORS, Types of: ASTABLE. A free-running multivibrator.

BISTABLE. A single trigger pulse switches conduction from one tube to the other.

FREE-RUNNING. Frequency of oscillation depending upon value of circuit components. Continuous oscillation.

MONOSTABLE. One trigger pulse is required to complete one cycle of operation.

ONE SHOT. Same as MONOSTABLE.

MUTUAL INDUCTANCE (M): When two coils are so located that magnetic flux of one coil will cause an emf in the other, there is mutual inductance.

NATURAL MAGNET: Magnets found in natural state in form of mineral called Magnetite.

NEGATIVE ION: Atom which has gained electrons and is negatively charged.

NETWORK: Two or more components connected in either series or parallel.

NEUTRON: Particle which is electrically neutral.

NICKEL CADMIUM CELL: Alkaline cell with paste electrolyte hermetically sealed.

NO LOAD VOLTAGE: Terminal voltage of battery or supply when no current is flowing in external circuit.

NONLINEAR DEVICE: Electronic device or component whose current-voltage relation is not a straight line.

NUCLEONICS: The branch of physics dealing with the science of small particles and the release of energy from the atom.

NUCLEUS: Core of the atom.

OHM: Electricity may or may not experience some difficulty or opposition as it flows through a circuit. This is called resistance and it is measured in ohms. Some materials have very high resistance; others allow electricity to flow easily. A circuit containing one ohm of resistance will allow one ampere of current to flow if one volt of electrical force is connected to it. Larger units of resistance may be written as:

KILOHM - one thousand ohms or 1,000 ohms.

MEGOHM - one million ohms or 1,000,000 ohms.

OHMMETER: Meter used to measure resistance in ohms.

OHM'S LAW: Mathematical relationship between current, voltage and resistance discovered by George Simon Ohm.

$$I = \frac{E}{R} \qquad E = IR \qquad R = \frac{E}{I}$$

OHMS PER VOLT: Unit of measurement of sensitivity in a meter.

OR CIRCUIT: A circuit with two or more inputs and one input signal must be present to produce an output signal.

OSCILLATOR: An electron tube generator of alternating current voltages.

OSCILLATORS, Types of: ARMSTRONG. An oscillator using tickler coil for feedback.

COLPITTS. An oscillator using split tank capacitor as feedback circuit.

CRYSTAL-CONTROLLED. An oscillator controlled by piezoelectric effect.

ELECTRON-COUPLED OSCILLATOR (ECO). Combination oscillator and power amplifier utilizing electron stream as coupling medium between grid and plate tank circuits.

HARTLEY. Oscillator using inductive coupling of tapped tank coil for feedback.

PUSH-PULL. Push-pull circuit utilizing interelectrode capacitance of each tube to feed back energy to grid circuit to sustain oscillations.

RC OSCILLATORS. Oscillators depending upon charge and discharge of capacitor in series with resistance.

OSCILLOSCOPE: Test instrument using cathode ray tube, permitting observation of signal.

OUTPUT: The energy from a circuit or device.

PARALLEL CIRCUIT: Circuit which contains two or more paths for electrons supplied by common voltage source.

PARALLEL RESONANCE: Parallel circuit of an inductor and capacitor at frequency when inductive and capacitive reactances are equal. Current in capacitive branch is 180 deg. out of phase with inductive current and their vector sum is zero.

PEAK: Maximum value of sine wave; the highest voltage current or power reached during a particular cycle or operating time.

PEAK INVERSE VOLTAGE: Value of voltage applied in reverse direction across diode.

PEAK INVERSE VOLTAGE RATING: The inverse voltage a diode will withstand without arcback.

PEAK-TO-PEAK: Measured value of sine wave from peak in positive direction to peak in negative direction.

PEAK VALUE: Maximum value of an alternating current or voltage.

PENTAGRID CONVERTER: Tube with five grids.

PENTAVALENT: Semiconductor impurity having five valence electrons. Donor impurities.

PENTODE: Electron tube with five elements including cathode, plate, control grid, screen grid and suppressor grid.

PERIOD: The time in seconds elapsed during one cycle. It is equal to one divided by frequency, $P = \frac{1}{f}$

PERMANENT MAGNET: Bars of steel and other substances which have been permanently magnetized.

PERMEABILITY (symbol μ): Relative ability of substance to conduct magnetic lines of force as compared with air.

PHASE: Relationship between two vectors in respect to angular displacement.

PHASE SPLITTER: Amplifier which produces two waves that have exactly opposite polarities from single input wave form.

PHOTODIODE: A PN junction diode which conducts upon exposure to light energy.

PHOTOELECTRIC CELL: A cell which produces an electric potential when exposed to light.

PHOTOELECTRIC EFFECT: The property of certain substances to emit electrons when subjected to light.

PHOTOELECTRONS: Electrons emitted as a result of light.

PHOTON: A discrete quantity of electromagnetic energy; a quantum.

PHOTOSENSITIVE: Characteristic of material which emits electrons from its surface when energized by light.

PHOTOTUBE: Vacuum tube employing photo sensitive material as its emitter or cathode.

PHOTOVOLTAIC: The generation of a voltage at the junction of two materials when exposed to light.

PIEZOELECTRIC EFFECT: Property of certain crystalline substances of changing shape when an emf is impressed upon crystal. Action is also reversible.

PITCH: Property of musical tone determined by its frequency.

PLANETARY ELECTRONS: Electrons considered in orbit around the nucleus of an atom.

PLATE: Anode of vacuum tube. Element in tube which attracts electrons.

PLUS (symbol +): Positive terminal or junction of circuit.

PM SPEAKER: Speaker employing permanent magnet as its field.

PN JUNCTION: The line of separation between N-type and P-type semiconductor materials.

POINT CONTACT DIODE: Diode consisting of point and a semiconductor crystal.

POLARITY: Property of device or circuit to have poles such as north and south or positive and negative.

POLARIZATION: Defect in cell caused by hydrogen bubbles surrounding positive electrode and effectively insulating it from chemical reation.

POLES: Number of poles in motor or generator field.

POSITIVE ION: Atom which has lost electrons and is positively charged.

POT: Abbreviation for potentiometer.

POWER: Rate of doing work. In dc circuits, $P = I \times E$.

POWER AMPLIFICATION: Ratio of output power to input grid driving power.

POWER SUPPLY: Electronic circuit designed to provide various ac and dc voltages for equipment operation. Circuit may include transformers, rectifiers, filters and regulators.

POWER TRANSISTOR: Transistors designed to deliver a specified output power level.

PREAMPLIFIER: Sensitive low-level amplifier with sufficient output to drive standard amplifier.

PREFIXES: A word united or joined to the beginning of another word to change its meaning. The new approved multiples are shown in the accompanying chart.

PREFIXES: Multiples

10^{12}	tera-	10^{-1}	deci-
10^{9}	giga-	10^{-2}	centi-
10^{6}	mega-	10^{-3}	milli-
10^{4}	myria-	10^{-6}	micro-
10^{3}	kilo-	10^{-9}	nano-
10^{2}	hekto-	10^{-12}	pico-
10	deka-		

PRIMARY CELL: Cell that cannot be recharged.

PRIMARY WINDING: Coil of transformer which received energy from ac source.

PROBE: A pointed metal end of a test lead, to contact specific points in a circuit to be measured.

PROTON: Positively charged particle.

PULSE: Sudden rise and fall of a voltage or current.

SPIKE. An unwanted pulse of relatively short duration superimposed on a main pulse.

SYNC. A pulse sent by a TV transmitter to synchronize the scanning of the receiver with the transmitter; a pulse used to maintain predetermined speed and/or phase relations.

PUSH-PULL AMPLIFIER: Two tubes used to amplify signal in such a manner that each tube amplifies one half-cycle of signal. Tubes operate 180 deg. out of phase.

Q: Letter representation for quantity of electricity (coulomb).

Q: Quality, figure of merit; ratio between energy stored in inductor during time magnetic field is being established to losses during same time. $Q = \frac{X_L}{R}$

QUIESCENT: At rest. Inactive.

RADIO FREQUENCY CHOKE (RFC): Coil which has high impedance to rf currents.

RADIO SPECTRUM: Division of electromagnetic spectrum used for radio.

RADIO WAVE: A complex electrostatic and electromagnetic field radiated from a transmitter antenna.

REACTANCE: INDUCTIVE (X_L). The opposition to the flow of an alternating current as the result of counter voltages induced in the coil by a varying current. It is measured in OHMS and is directly proportional to the frequency of the applied voltage and the value of the coil in henrys.

CAPACITIVE (X_C). The opposition to the flow of an alternating current as the result of counter voltages stored in the capacitor when a varying voltage is applied. It is measured in OHMS and is inversely proportional to the frequency of applied voltage and the value of the capacitor in farads.

RECIPROCAL: Reciprocal of number is one divided by the number.

RECTIFIER: Component or device used to convert ac into a pulsating dc.

REGULATION: Voltage change that takes place in output of generator or power supply when load is changed.

REGULATION, PERCENTAGE OF: Percentage of change in voltage from no-load to full-load in respect to full-load voltage.

$$\text{Expressed as:} \quad \frac{E_{no\ load} - E_{full\ load}}{E_{full\ load}} \times 100$$

REJECT CIRCUIT: Parallel tuned circuit at resonance. Rejects signals at resonant frequency.

RELAY: Magnetic switch.

RELAXATION OSCILLATOR: Nonsinusoidal oscillator whose frequency depends upon time required to charge or discharge capacitor through resistor.

REPULSION-START MOTOR: A motor which develops starting torque by interaction of rotor currents and single-phase stator field.

RESIDUAL MAGNETISM: Magnetism remaining in material after magnetizing force is removed.

RESISTANCE: Quality of electric circuit that opposes flow of current through it.

RESONANT FREQUENCY: Frequency at which tuned circuit oscillates. (See TUNED CIRCUIT.)

RETENTIVITY: Ability of material to retain magnetism after magnetizing force is removed.

RETRACE: Process of returning scanning beam to starting point after one line is scanned.

RIPPLE VOLTAGE: AC component of DC output of power supply due to insufficient filtering.

RMS VALUE: ROOT-MEAN-SQUARE value. The same as effective value. $(.707 \times E_{peak})$

ROTOR: Rotating part of an ac generator.

SAWTOOTH GENERATOR: Electron tube oscillator producing sawtooth wave.

SAWTOOTH WAVE: Wave shaped like the teeth of a saw.

SCHEMATIC: Diagram of electronic circuit showing electrical connections and identification of various components.

SCREEN GRID: Second grid in electron tube between grid and plate, to reduce interelectrode capacitance.

SECONDARY CELL: Cell that can be recharged by reversing chemical action with electric current.

SECONDARY EMISSION: Emission of electrons as result of electrons striking plate of electron tube.

SECONDARY WINDING: Coil which receives energy from primary winding by mutual induction and delivers energy to load.

SELECTIVITY: Relative ability of receiver to select desired signal while rejecting all others.

SELF-INDUCTANCE: Emf is self-induced when it is induced in conductor carrying current.

SEMICONDUCTOR: Conductor with resistivity somewhere in range between conductors and insulators.

SEMICONDUCTOR, N TYPE: Semiconductor which uses electrons as major carrier.

SEMICONDUCTOR, P TYPE: Semiconductor which uses holes as major carrier.

SENSITIVITY: Ability of circuit to respond to small signal voltages.

SENSITIVITY OF METER: Indication of loading effect of meter. Resistance of moving coil and multiplier divided by voltage for full scale deflection. Sensitivity equals one divided by current required for full scale deflection. Ex: A 100 μA meter movement has sensitivity of: $\frac{1}{.0001}$ or 10,000 ohms/volt.

SERIES CIRCUIT: Circuit which contains only one possible path for electrons through circuit.

SERIES PARALLEL: Groups of series cells with output terminals

connected in parallel.

SERIES RESONANCE: Series circuit of inductor, a capacitor and resistor at a frequency when inductive and capacitive reactances are equal and cancelling. Circuit appears as pure resistance and has minimum impedance.

SHIELD: Partition or enclosure around components in circuit to minimize effects of stray magnetic and radio frequency fields.

SHORT CIRCUIT: Direct connection across source which provides zero resistance path for current.

SHUNT: To connect across or parallel with circuit or component.

SHUNT: Parallel resistor to conduct excess current around meter moving coil. Shunts are used to increase range of meter.

SIDEBANDS: Frequencies above and below carrier frequency as result of modulation.

LOWER. Frequencies equal to difference between carrier and modulating frequencies.

UPPER. Frequencies equal to carrier plus modulating frequencies.

SIGNAL: The intelligence, message or effect to be sent over a communications system; an electrical wave corresponding to intelligence.

SINE WAVE: A wave form of a single frequency alternating current. A wave whose displacement is the sine of an angle proportional to time or distance.

SINGLE ENDED AMPLIFIER: An amplifier whose final power stage is a single vacuum tube or transistor.

SINGLE PHASE MOTOR: Motor which operates on single phase alternating current.

SLIP RINGS: Metal rings connected to rotating armature windings in generator. Brushes sliding on these rings provide connections for external circuit.

SOLENOID: Coil of wire carrying electric current possessing characteristics of magnet.

SPACE CHARGE: Cloud of electrons around cathode of an electron tube.

SPECIFIC GRAVITY: Weight of liquid in reference to water which is assigned value of 1.0.

SPLIT PHASE MOTOR: Single phase induction motor which develops starting torque by phase displacement between field windings.

SQUIRREL CAGE ROTOR: Rotor used in an induction motor made of bars placed in slots of rotor core and all joined together at ends.

STABILITY: The ability to stay on a given frequency or in a given state without undesired variation.

STAGE: Section of an electronic circuit, usually containing one electron tube and associated components.

STATIC CHARGE: Charge on body either negative or positive.

STATIC ELECTRICITY: Electricity at rest as opposed to electric current.

STATOR: Stationary coils of an ac generator.

STEADY STATE: Fixed nonvarying condition.

STORAGE BATTERY: Common name for lead-acid battery used in automotive equipment.

STYLUS: Phonograph needle or jewel, which follows grooves in a record.

SUBHARMONIC: A frequency below harmonic, usually fractional part of fundamental frequency.

SUBSONIC: A frequency below the audio frequency range; infrasonic.

SUPERSONIC: Frequencies above audio frequency range.

SUPERHETERODYNE: Radio receiver in which incoming signal is converted to fixed intermediate frequency before detecting audio signal component.

SUPPRESSOR GRID: Third grid in electron tube, between screen grid and plate, to repel or suppress secondary electrons from plate.

SWEEP CIRCUIT: Periodic varying voltage applied to deflection circuits of cathode ray tube to move electron beam at linear rate.

SWITCH: Device for directing or controlling current flow in circuit.

SYNC PULSE: Abbreviation for synchronization pulse, used for triggering an oscillator or circuit.

TANK CIRCUIT: Parallel resonant circuit.

TAP: Connection made to coil at point other than its terminals.

TAPE TRANSPORT: The driving mechanism and reels of a tape recorder.

TELEVISION: Method of transmitting and receiving visual scene by radio broadcasting.

TELEVISION CHANNEL: Allocation in frequency spectrum of 6 MHz assigned to each television station for transmission of picture and sound information.

TETRODE: Electron tube with four elements including cathode, plate, control grid and screen grid.

THERMISTOR: Semiconductor device which changes resistivity with change in temperature.

THETA (θ): Angle of rotation of vector representing selected instants at which sine wave is plotted. Angular displacement between two vectors.

THREE-PHASE ALTERNATING CURRENT: Combination of three alternating currents having their voltages displaced by 120 deg. or one-third cycle.

TICKLER: Coil used to feed back energy from output to input circuit.

TIME CONSTANT (RC): Time period required for the voltage of a capacitor in an RC circuit to increase to 63.2% of maximum value or decrease to 36.7% of maximum value.

TONE CONTROL: Adjustable filter network to emphasize either high or low frequencies in output of audio amplifier.

TRANSCEIVER: A combined transmitter and receiver.

TRANSDUCER: Device by which one form of energy may be converted to another form, such as electrical, mechanical or acoustical.

TRANSFORMER: Device which transfers energy from one circuit to another by electromagnetic induction.

TRANSFORMERS, Types: ISOLATION. Transformer with one-to-one turns ratio.

STEP-DOWN. Transformer with turns ratio grater than one. The output voltage is less than input voltage.

STEP-UP. Transformer with turns ratio of less than one. Output voltage is greater than input voltage.

TRANSIENT RESPONSE: Response to momentary signal or force.

TRANSISTOR: Semiconductor device derived from two words, transfer and resistor.

TRANSISTOR SOCKET: A small device in which the leads of a transistor are placed to provide ease in connection to circuit and to permit replacement of transistor.

TRANSMISSION LINE: Wire or wires used to conduct or guide electrical energy.

TRANSMITTER: Device for converting intelligence into electrical impulses for transmission through lines or through space from radiating antenna.

TRIMPOT: The trade name for a precision variable resistor manufactured by Bourns.

TRIODE: Three-element vacuum tube, consisting of cathode, grid and plate.

TRIVALENT: Semiconductor impurity having three valence electrons. Acceptor impurity.

TRUE POWER: Actual power absorbed in circuit.

TUNE: The process of bringing a circuit into resonance by adjusting one or more variable components.

TUNED AMPLIFIER: Amplifier employing tuned circuit for input and/or output coupling.

TUNED CIRCUIT: Circuit containing capacitance, inductance and resistance in series or parallel, which when energized at specific frequency known as its resonant frequency, an interchange of energy occurs between coil and capacitor.

TURNS RATIO: Ratio of number of turns of primary winding of transformer to number of turns of secondary winding.

TWEETER: A high frequency speaker.

UNIT MEASUREMENT: Standard increments of measurement or quantity used as means of comparison to other quantities.

UNIVERSAL MOTOR: Series ac motor which operates also on dc. Fractional horsepower ac-dc motor.

UNIVERSAL TIME CONSTANT CHART: A graph with curves representing growth and decay of voltages and currents in RC and RL circuits.

VALENCE: The chemical combining ability of an element in reference to hydrogen. The capacity of an atom to combine with other atoms to form molecules.

VECTOR: Straight line drawn to scale showing direction and magnitude of a force.

VECTOR DIAGRAM: Diagram showing direction and magnitude of several forces, such as voltage and current, resistance, reactance and impedance.

VOICE COIL: Small coil attached to speaker cone, to which signal is applied. Reaction between field of voice coil and fixed magnetic field causes mechanical movement of cone.

VOLT: A unit used to describe a difference between two levels of electrical force. This force will cause the flow of electricity between the two levels if connected by a circuit or wires. A flashlight cell will be rated at 1.5 volts. The automotive battery has a 12 volt difference between its terminals. The electricity we use at home is about 117 volts. Very often units smaller or larger than a volt are used for convenience. Study the following terms.

MEGAVOLT equals one million volts or 1,000,000 volts.

KILOVOLT equals one thousand volts or 1,000 volts.

MILLIVOLT equals one thousandth of a volt or .001 volt.

MICROVOLT equals one millionth of a volt or .000001 volt.

VOLTAGE DIVIDER: Tapped resistor or series resistors across source voltage to produce multiple voltages.

VOLTAGE DOUBLER: Rectifier circuit which produces double the input voltage.

VOLTAGE DROP: Voltage measured across resistor. Voltage drop is equal to product of current times resistance in ohms. E = IR.

VOLTAIC CELL: Cell produced by suspending two dissimilar elements in acid solution. Potential difference is developed by chemical action.

VOLTMETER: Meter used to measure voltage.

VOM: A common test instrument which combines a voltmeter, ohmmeter and milliammeter in one case.

VTVM: Vacuum Tube Voltmeter.

VU: Number numerically equal to number of decibels above or below reference volume level. Zero Vu represents power level of one milliwatt dissipated in 600 ohm load or voltage of .7746 volts.

WATT: This unit is a measurement of power used in an electrical circuit. You already know about mechanical power. Your car engine may be rated at two or three hundred horsepower. Power is the rate of doing work. In electricity work is done when an electrical current overcomes the resistance in a circuit. The result is heat which is radiated out to the surrounding air. Your electric toaster at home may be rated at 550 watts. You use 75 and 100 watt light bulbs. Other larger and smaller units are:

MEGAWATT   - one million watts or 1,000,000 watts.

KILOWATT   - one thousand watts or 1,000 watts.

MILLIWATT   - one thousandth of a watt or .001 watt.

MICROWATT   - one millionth of a watt or .000001 watt.

WATT-HOUR: This term describes the use of electrical power. It is also the unit by which you purchase power from the power company. If you use one watt of power for one hour, your power consumption would be one watt-hour. This unit is very small. Therefore the meter on the outside of your home reads in KILOWATT-HOURS or KWH. If you consume one thousand watts per hour, you use one KWH. A one hundred watt lamp burning for ten hours would use one KWH. The power consumed by a circuit or device can always be found by the formula, VOLTS times AMPERES equals WATTS. Multiply your answer by time in hours and you have watt-hours.

WATTLESS POWER: Power not consumed in an ac circuit due to reactance.

WATTMETER: Meter used to measure power in watts.

WAVE: A disturbance in a medium which is a function of time and space or both. Energy may be transmitted by waves. Ex: Audio and radio waves.

WAVE FORM: The shape of a wave derived from plotting its instantaneous values during a cycle against time.

WAVELENGTH: Distance between point on loop of wave to corresponding point on adjacent wave.

WHEATSTONE BRIDGE: Bridge circuit used for precision measurement of resistor.

WOOFER: A large speaker designed for reproduction of low frequency sounds. It is used in high fidelity amplifier systems.

WORK: When a force moves through a distance, work is done. It is measured in foot-pounds and Work = Force x Distance.

X-AXIS: The horizontal axis of a graph.

Y-AXIS: The vertical axis of a graph.

ZENER DIODE: A silicon diode which makes use of the breakdown properties of a PN junction. If a reverse voltage across the diode is progressively increased, a point will be reached when the current will greatly increase beyond its normal cutoff value. This voltage point is called the Zener voltage.

ZENER VOLTAGE: The reverse voltage at which the breakdown occurs in a Zener diode.

INDEX